Advances in Modern Blind Signal Separation Algorithms

Theory and Applications

Synthesis Lectures on Algorithms and Software in Engineering

Editor
Andreas S. Spanias, *Arizona State University*

Advances in Modern Blind Signal Separation Algorithms: Theory and Applications
Kostas Kokkinakis and Philipos C. Loizou

ISBN: 978-3-031-00384-4   paperback
ISBN: 978-3-031-01512-0   ebook

DOI 10.1007/978-3-031-01512-0

A Publication in the Springer series
SYNTHESIS LECTURES ON ALGORITHMS AND SOFTWARE IN ENGINEERING

Lecture #6
Series Editor: Andreas S. Spanias, *Arizona State University*
Series ISSN
Synthesis Lectures on Algorithms and Software in Engineering
Print 1938-1727   Electronic 1938-1735

Advances in Modern Blind Signal Separation Algorithms

Theory and Applications

Kostas Kokkinakis and Philipos C. Loizou
University of Texas at Dallas

SYNTHESIS LECTURES ON ALGORITHMS AND SOFTWARE IN ENGINEERING #6

ABSTRACT

With human-computer interactions and hands-free communications becoming overwhelmingly important in the new millennium, recent research efforts have been increasingly focusing on state-of-the-art multi-microphone signal processing solutions to improve speech intelligibility in adverse environments. One such prominent statistical signal processing technique is *blind signal separation* (BSS). BSS was first introduced in the early 1990s and quickly emerged as an area of intense research activity showing huge potential in numerous applications. BSS comprises the task of 'blindly' recovering a set of unknown signals, the so-called sources from their observed mixtures, based on very little to almost no prior knowledge about the source characteristics or the mixing structure. The goal of BSS is to process multi-sensory observations of an inaccessible set of signals in a manner that reveals their individual (and original) form, by exploiting the spatial and temporal diversity, readily accessible through a multi-microphone configuration. Proceeding blindly exhibits a number of advantages, since assumptions about the room configuration and the source-to-sensor geometry can be relaxed without affecting overall efficiency.

This booklet investigates one of the most commercially attractive applications of BSS, which is the simultaneous recovery of signals inside a reverberant (naturally echoing) environment, using two (or more) microphones. In this paradigm, each microphone captures not only the direct contributions from each source, but also several reflected copies of the original signals at different propagation delays. These recordings are referred to as the convolutive mixtures of the original sources. The goal of this booklet in the lecture series is to provide insight on recent advances in algorithms, which are ideally suited for blind signal separation of convolutive speech mixtures. More importantly, specific emphasis is given in practical applications of the developed BSS algorithms associated with real-life scenarios. The developed algorithms are put in the context of modern DSP devices, such as hearing aids and cochlear implants, where design requirements dictate low power consumption and call for portability and compact size. Along these lines, this booklet focuses on modern BSS algorithms which address (1) the limited amount of processing power and (2) the small number of microphones available to the end-user.

KEYWORDS

acoustics, cocktail-party effect, speech enhancement, blind signal separation, noise reduction, filters, algorithms, multiple microphones, convolutive speech mixtures, hearing aids, cochlear implants

Contents

Acknowledgments

A large part of this book is derived from the research activities of the two authors over the past five years. Most of this original work, as well as writing this book have been conducted in the Erik Jonsson School of Engineering and Computer Science at the University of Texas at Dallas, Richardson, TX, USA. Much of this research has been made possible by funding from the National Institute of Deafness and other Communication Disorders (NIDCD) of the National Institutes of Health (NIH), which is gratefully acknowledged. The authors would also like to thank the cochlear implant patients for their time and dedication during their participation in our clinical studies, as well as the help of Cochlear Limited and Jan Wouters of ESAT of K.U. Leuven. Lastly, and most importantly, we would also like to take this opportunity and thank our families for their support, understanding and encouragement during the writing of this book.

Kostas Kokkinakis and Philipos C. Loizou
Richardson, TX

Preface

Blind signal separation (BSS) is a prominent statistical signal processing strategy, which seeks to recover the individual contributions of a set of unknown but statistically independent physical sources. BSS relies only on measurements observed at each input channel, while it assumes little to almost no prior knowledge regarding the source-to-sensor geometry or the source signals themselves.

In Chapter 1, we present necessary background material. The chapter begins with a mathematical description of the blind signal separation framework and concludes with a thorough review of the generative models for both instantaneous and convolutive mixtures.

In Chapter 2, we elucidate on the most significant theoretical aspects that underpin the development of convolutive blind signal separation strategies and review some of the most promising BSS algorithms and how they relate to speech processing applications.

In Chapter 3, we focus on the application of blind signal processing strategies to noise reduction for the hearing-impaired. To enhance the ability of hearing-impaired individuals to communicate better in noise, we describe noise reduction processing strategies based on BSS, which exploit an array of two or more microphones.

In Chapter 4, we summarize the main issues which have been discussed in this booklet and furthermore we outline the direction that future research in this area could follow.

Kostas Kokkinakis   University of Texas at Dallas
Philipos C. Loizou   January 31, 2010

CHAPTER 1

Fundamentals of blind signal separation

1.1 INTRODUCTION

Consider the familiar scenario in which multiple sounds emitted from a number of different sources (e.g., other people) are perceived, segregated and recognized on demand. Despite the fact that all these individual sounds generated are summed up into a single acoustic waveform, the human brain possesses the seemingly basic, yet remarkable ability to segregate and subsequently focus on a single source of speech within a mixture of two or more sounds, which may include the presence of background noise, as well as a number of other competing voices (e.g., see Bronkhorst, 2000; Alain *et al.*, 2005). The essence of this phenomenon, which is usually referred to as the 'cocktail-party problem' — a definition which first appeared in the literature by Cherry (1953) and Cherry and Taylor (1954) — can be formulated as a deceptively simple question: "How can one recognize what another person is saying when other individuals are speaking at the same time?"

Finding answers to this question has been an important goal of human hearing research for several decades and over the years many attempts have concentrated on deciphering how the brain operates in these scenarios. Yet, even today most of the high-level perceptual mechanisms responsible for the completion of this task remain unknown. Comparably challenging are situations when sound is captured with the aid of a digital system consisting of an array of sensors (e.g., microphones). In this case, conventional systems, unlike humans, are unable to isolate the individual sources and hence only treat respective inputs as a single sound source. Early attempts to emulate the human listening perception process within the context of an intelligent system, have led to the development of new techniques, like computational auditory scene analysis (CASA) (Weintraub, 1985) and blind signal separation (BSS) (Jutten and Hérault, 1991). In its most basic form, CASA aims at extracting individual elements, normally referred to as auditory objects, by utilizing some form of perceptual grouping based on psychoacoustic principles (e.g., see Brown and Cooke, 1994; Wang and Brown, 2006). In this respect, most of the current CASA models make little to no use of world knowledge or contextual information to assist in the process of scene analysis and therefore are more concerned with the detection and outlining of sound objects, rather than the exact reconstruction of original sounds, which is not always feasible.

As opposed to CASA, which draws its inspiration mainly from the psychoacoustic aspects of human auditory organization, BSS requires no explicit modeling of the underlying auditory phenomena. In contrast to CASA, BSS is a well-defined and understood operation, and most importantly, one governed solely by straightforward statistical mathematical principles. As a result, since the early 1990s when it was first introduced, BSS has quickly emerged as an area of intensely growing research activity with potential use in numerous practical applications.

BSS comprises the task of blindly recovering a set of unknown original signals, the so-called sources from their observed mixtures, based entirely on little to no prior knowledge about the source characteristics (except that at most one can be Gaussian) or the mixing structure itself. This total lack of any *a priori* knowledge regarding the origin of the mixtures, however, is compensated well by the statistically strong yet often physically plausible assumption of statistical independence amongst all sources (Comon, 1994). On this basis, the principal aim of BSS is to process multisensory observations of an inaccessible set of signals (sources) in a manner that reveals their individual (and original) form. More importantly, in doing so, BSS is based exclusively on the premise that the retrieved signal components must be mutually statistically independent or in fact as independent as possible [1]. Normally, by following such a strategy any hidden or unknown original source signal can be retrieved well, by simply reconstructing the independent components from a set of available mixtures.

1.2   APPLICATIONS

In principle, proceeding blindly exhibits a number of advantages, mainly due to the fact that requirements about some prior knowledge of the mixing process (e.g., room configuration and source-to-sensor geometry), as well as information on the statistical profile of the sources (source prior) can be relaxed sufficiently without hindering the ability to achieve separation. This task of performing some sort of blind processing to recover a set of unknown time series in order to extract useful streams of information, arises in a handful of applications, with many practical examples featuring prominently in the related fields of acoustics and speech processing.

Nonetheless, being at the intersection of numerous engineering disciplines and an important application domain and research focus of several scientific communities, which include machine learning, neural networks, basic information theory, as well as statistical signal processing, the use of BSS has also been proven beneficial in a wide variety of other areas, such as the ones described below.

Acoustics. Cross-talk removal, speech separation, auditory perception, scene analysis, coding, recognition, synthesis and segmentation, psychoacoustics, reverberation, echo and noise suppression and cancellation, signal enhancement, automatic speech recognition (ASR) in reverberant and noisy acoustical settings. Potential uses in mobile telephony, hands-free devices, human-machine interfaces (HMIs), hearing aids, cochlear implants, airport surveillance, automobiles and aircraft cockpit environments (e.g., see Chaumette *et al.*, 1993; Omologo *et al.*, 1998; Visser *et al.*, 2003; Kokkinakis and Loizou, 2008).

---

[1] In this context, BSS is an equivalent process to independent component analysis (ICA), which by definition searches for a linear transformation that can minimize the statistical dependence between its components (Comon, 1994).

Biomedical processing. Non-invasive separation of fetal from maternal electrocardiograms (ECGs), enhancement, decomposition, artifact removal of electroencephalograms (EEGs), magnetoencephalograms (MEGs) and electromyograms (EMGs) (e.g., see Makeig *et al.*, 1996; Zarzoso and Nandi, 2001).

Digital communications. Smart antennae design, direction of arrival (DOA) estimation, adaptive beamforming, multichannel equalization, multiuser separation. Potential uses in wireless communication schemes, mobile radio and telephony and satellite telecommunication systems (e.g., see Godard, 1980; Tugnait, 1994; Talwar *et al.*, 1996).

Financial applications. Econometrics, analysis of financial time series (e.g., currency exchange rates, cashflow data), stock prediction, minimization of investment risks (e.g., see Back and Weigend, 1997; Cheung and Xu, 2001).

Geophysics. Geophysical exploration, seismic monitoring, data analysis, signal separation and prediction (e.g., see Walden, 1985; Thirion *et al.*, 1996).

Image processing. Image and video analysis, denoising, compression, restoration and reconstruction, computer vision and tomography, visual scene analysis, motion detection and tracking, medical imaging (e.g., see Karhunen *et al.*, 1997; Tonazzini *et al.*, 2004).

Statistics. Factor analysis, data mining, feature extraction and classification, Bayesian modeling, information theory (Lewicki, 1994).

1.3   INSTANTANEOUS MIXING MODEL

In the **instantaneous** BSS mixing model, we may assume that the original (source) signals are time aligned and that they arrive at the microphones (sensors) simultaneously without any time delays. This being one of the simplest approximations possible, assumes an instantaneous propagation with no dispersive (multipath) effects and in essence mixing coefficients, which are just scaling factors or as otherwise referred to as a memoryless channel (Pope and Bogner, 1996a). For example, this scenario holds true when the acoustic impulse responses between the respective sources and sensors are composed by just a single coefficient, thus having a structure roughly resembling the one shown in Figure 1.3 (b) (see Section 1.7). In this case, the observed signals are said to simply form **instantaneous mixtures** of the independent sources, with the elements of the unobservable mixing matrix taking the form:

$$h_{ij}(k) = a_{ij}\, \delta(k) \tag{1.1}$$

where $\delta(k)$ is the *Kronecker delta* defined as:

$$\delta(k) = \begin{cases} 1, & k = 0 \\ 0, & k \neq 0 \end{cases} \tag{1.2}$$

This yields the so-called linear **instantaneous (or spatial-only) BSS mixing model**, which can be written as:

$$x_i(t) = \sum_{j=1}^{n} a_{ij} s_j(t), \quad i = 1, 2, \ldots, m \tag{1.3}$$

or, alternatively, in matrix form:

$$\begin{pmatrix} x_1(t) \\ \vdots \\ x_m(t) \end{pmatrix} = \begin{pmatrix} a_{11} & \cdots & a_{1n} \\ \vdots & \ddots & \vdots \\ a_{m1} & \cdots & a_{mn} \end{pmatrix} \begin{pmatrix} s_1(t) \\ \vdots \\ s_n(t) \end{pmatrix} \Rightarrow$$

$$\Rightarrow \mathbf{x}(t) = A\,\mathbf{s}(t) \tag{1.4}$$

Based on this type of formulation, the most widely used technique to perform BSS is **independent component analysis (ICA)** (Comon, 1994). The explicit mathematical assumption of a linear, instantaneous (or static) and time-invariant mixing stage implies that the underlying propagation effects encountered in typical acoustic scenarios are assumed not to occur (or be negligible). Arguably, such a simplification may be unrealistic especially in real echoic situations, since it fails to accommodate cases in which the transmission channels are dynamically changing. In effect, the practical use of solutions based on such a model are inherently limited.

1.4   ASSUMPTIONS

In principle, even the simplified model of Eq. (1.3) would certainly be ill-posed if no further assumptions were made about the characteristics of the system. These hypotheses can be divided into three categories, depending on whether they refer to the mixing matrix, the sources or the noise signals. The standard hypotheses are the following:

A1. The number of sensors is equal to or greater than the number of the original sources, such that $m \geqslant n$.

A2. At each time instant t, the components of the source vector $\mathbf{s}(t)$ are mutually statistically independent.

A3. At most one source can follow a Gaussian distribution.

A4. The components of the source vector $\mathbf{s}(t)$ are zero-mean and unit-variance processes.

A5. There exists no additive noise at the sensors.

Assumption A1 states that the number of sensors must be at least equal to the number of sources. This condition guarantees that the mixing matrix is full column rank[2] and hence, invertible. In general, one can recover, at most, as many sources as sensors, making $m = n$ and hence the assumption of a square mixing matric, one of the most widely assumed conditions in theoretical BSS formulations. However, in practical problems, it may often be unreasonable to assume that the number of (independent) sources is always equal to the number of observed measurements. Based on the number of sources and sensors, we can distinguish between three important cases:

- When $n = m$, the number of independent sources is equal to the number of the measurements observed at the sensors. This condition implies a **fully determined** system.

- When $n < m$, the number of independent sources is less than the number of the measurements observed at the sensors. This describes the **overdetermined** BSS scenario, which in some cases can also be referred to as the problem of *undercomplete* bases (Amari, 1999; Cichocki and Amari, 2002).

- When $n > m$, the number of measurements observed at the sensors is less than the number of independent sources. This is often called **underdetermined** BSS (Lee *et al.*, 1999; Araki *et al.*, 2007). Note that underdetermined BSS is also known as the *overcomplete* bases problem.

Assumption A2 is the essence of blind signal separation models. Statistical independence amongst the different sources is a statistically strong, but often physically plausible assumption[3]. The spatial whiteness or spatial independence of the sources is a key assumption for BSS. It implies that the source joint probability density function (PDF) denoted by $p_s(\mathbf{s}(t))$ is equal to the product of the source marginal independent PDFs, meaning that it can be factorized such that:

$$p_s(\mathbf{s}(t)) = \prod_{i=1}^{n} p_{s_i}(s_i(t)) \tag{1.5}$$

Any set of random variables that obey Eq. (1.5) are said to be pairwise statistically independent. Assumption A3 unveils a fundamental limitation of ICA. In principle, due to the central limit

---

[2]In the context of linear algebra, the column rank of a matrix is defined as the maximum number of columns, which are linearly independent. For $A \in \mathbb{R}^{m \times n}$ term full column rank implies rank equal to n.

[3]BSS exploits the fact that two or more signals, such as speech emitted from different physical sources, e.g., two different talkers are mutually statistically independent (Comon, 1994). Put simply, two or more speech signals are said to be independent of each other, if and only if the amplitude of one signal provides no information with respect to the amplitude of the other, at any given time.

theorem (CLT), any linear combination of independent and identically distributed (i.i.d) Gaussian random variables is Gaussian itself. This implies that for Gaussian source signals, the original and mixed distributions are identical. Hence, when confronted with a pair of Gaussian independent components, it is in general impossible[4] to recover the original source signals from their mixtures, without resorting to some extra knowledge about the sources themselves. Assumption A4, stipulates the so-called zero-mean and unit-variance normalization convention, with regard to the amplitude of the (unknown) source signals. Although such an assumption is almost always taken for granted, it is not a necessary condition for BSS, since even non-zero mean sources with different powers can still be accommodated for, through an extra constant factor in the model (Cardoso, 1998). Assumption A5, essentially restricts the applicability of BSS to noiseless scenarios only. In general, the estimation of source estimates in the presence of noise is a rather difficult task and requires a number of additional assumptions to be made. Therefore, the majority of BSS models often stipulate that there is no additive noise components present at the sensors.

Despite the general applicability and apparent usefulness of the BSS model, there are cases where it is not always possible uniquely identify the original source signals. Inevitable ambiguities in blind separation are (1) scaling and (2) permutation. Consequently, source recovery and mixing matrix identification can only be accomplished up to certain indeterminacies (Hyvärinen et al., 2001):

Scaling. The variances (or energies) of the recovered sources cannot be determined exactly. In fact, since both the mixing matrix and the original sources are unknown, any scalar multiplier can be exchanged between each source and the corresponding column of the mixing matrix without modifying the observations at all.

Permutation. The order of the recovered sources cannot be determined exactly. Essentially, the order of the sources is itself arbitrary, since re-ordering the sources and the columns of the mixing matrix accordingly, leaves the observations unaltered. This suggests that the source signals can be at best recovered up to a permutation.

Therefore, the goal of the BSS task in the case of instantaneous mixtures, requires the estimation of a linear transformation W, in order to extract the original source signals, such that:

$$\mathbf{u} = \mathbf{W}\mathbf{x} = \mathbf{W}\mathbf{A}\,\mathbf{s} = \boldsymbol{P}\boldsymbol{D}\,\mathbf{s} \qquad (1.6)$$

in which case the combination of the mixing and unmixing or the so-called global system matrix can be written as:

$$\boldsymbol{G} = \boldsymbol{P}\boldsymbol{D} \qquad (1.7)$$

[4]Note, however, that it is still possible to separate spatially uncorrelated but temporally correlated Gaussian sources under specific conditions, by jointly exploiting their non-stationarity and (second-order) temporal statistics.

In this case, P is a permutation matrix of size $(n \times n)$ and matrix D is a non-singular diagonal matrix of the same dimensions. Note that the former matrix represents the re-ordering or labeling of the source signals, while the latter is concerned with any possible scaling ambiguities. Although, scaling and permutation uncertainties appear to be rather severe limitations; in general, they do not diminish or limit the applicability of BSS. This is due to the fact that the most relevant information about the original sources is contained in the waveforms of the signals and not in their respective amplitudes or the order, in which they are arranged at the output of the unmixing system. The concept of invariance regarding the BSS model to scaling and permutations was further strengthened by Tong *et al.* (1993), who showed that any transformation performed on independent signals also produces independent signals and is a transformation within the waveform (preserving) equivalence class if and only if, at most, one of the components of the source signals is Gaussian in nature. In any case, any recovered source vector obeying the form shown in Eq. (1.7) can be regarded as an exact signal copy of the original sources.

1.5   INDEPENDENT & PRINCIPAL COMPONENT ANALYSIS

The simplest approximation of the BSS model in which the mixing coefficients are approximated with scaling factors (instantaneous mixing) is directly related to the concept of **independent component analysis (ICA)**. ICA has been extensively studied in the literature with the earliest methods tracing back to the pioneering work of Cardoso (1989) and Jutten and Hérault (1991). However, the first ever formal definition of ICA came from the influential work of Comon (1994). In his seminal 1994 paper, he coined the problem in terms of statistical independence and progressed it further by forming its cardinal mathematical formulation and relating it to the widely-employed concept of BSS. As formulated by Comon, performing ICA on a random vector consists of searching for a linear non-orthogonal transformation that maximizes the statistical independence (up to a given order) between its components.

In most cases, this can be facilitated by resorting to *higher-order statistics* (HOS)[5] either implicitly or explicitly. In principle, ICA can be carried out by invoking the use of higher-order moments or cumulants (Cardoso and Laheld, 1996) and form specific contast functions that need to be optimized (Comon, 1994). In any case, ICA will always aim to reducing any higher-order statistical dependencies between a set of multivariate data. For this purpose, higher-order statistics can be explored extensively, in cases where the source components are assumed to be non-Gaussian processes, as well as, spatially independent at each time instant. For a nice overview of different techniques for blind source separation in the case of linear instantaneous mixtures, based entirely on HOS, as well as for a more complete set of references on the subject, the reader is referred to Nandi (1999).

---

[5] The term HOS refers to moments and cumulants of order greater than two. On most occasions, these higher-order moments and cumulants reveal more useful information than that extracted from only the first- and second-order statistics of the data, namely the mean and variance.

In practice, higher-order statistical independence can be seen as an extension (and a special case) of second-order independence (or uncorrelatedness) and, therefore, of **principal component analysis (PCA)** (Hyvärinen *et al.*, 2001). In fact, one may say that the property of independence is stronger, due to the fact that all processes that are independent are also uncorrelated. However, processes which are uncorrelated are not necessarily independent[6]. PCA, also known as the Karhunen-Loève transform (KLT) or data whitening (sphering), is a widely applicable statistical technique, which in many ways can be considered as the predecessor of ICA since it also seeks the removal of dependence from a set of data, albeit only up to second-order. Resorting only to correlations between data, PCA is carried out by employing conventional techniques based on *second-order statistics* (SOS) only. By applying PCA, data are then projected into their dominant principal components dimensional space as determined by the directions of maximum variance and hence a linear transformation Q is obtained, such that the outputs obtained are orthogonal or uncorrelated. Therefore, when given a random vector $\mathbf{x}(t) = [x_1(t), \ldots, x_m(t)]^T$, the aim of PCA is to find such a linear transformation Q, which transforms $\mathbf{x}(t)$ into another zero-mean and unit-variance vector $\mathbf{z}(t) = [z_1(t), \ldots, z_m(t)]^T \in \mathbb{R}^m$ such that (after dropping the time-index t for convenience):

$$\mathbf{z} = Q\mathbf{x} \tag{1.8}$$

with the spatial covariance matrix R_z being equal to:

$$R_z = \mathrm{E}[\mathbf{z}\mathbf{z}^T] = I_m \tag{1.9}$$

where $\mathrm{E}[\cdot]$ and $(\cdot)^T$ represent the expectation and transpose operator, respectively. In practice, to perform PCA one may consider carrying out the eigenvalue decomposition (EVD) of the sensor output covariance matrix $R_x = \mathrm{E}[\mathbf{x}\mathbf{x}^T]$, yielding:

$$Q = \Lambda^{-\frac{1}{2}} \Gamma^T \tag{1.10}$$

in which the orthogonal matrix Γ contains the eigenvectors of R_x and Λ contains its eigenvalues. One can easily verify that R_z is the identity matrix under such transformation.

1.6   ILLUSTRATION OF PCA AND ICA

In this section, we attempt to describe intuitively how ICA operates on a set of mixture data by resorting to the following example. Consider the pair of signals $\mathbf{s}(k) = [s_1(t), s_2(t)]^T$, which represent a set of statistically independent zero-mean and unit-variance sources with uniform distributions, such that:

$$p_{s_i}(s_i) = \begin{cases} \dfrac{1}{2\sqrt{5}}, & s_i \in \left[-\frac{1}{\sqrt{5}}, \frac{1}{\sqrt{5}}\right] \\[2ex] 0, & \text{elsewhere} \end{cases} \tag{1.11}$$

---

[6]The only case where uncorrelatedness implies independence is when the underlying source distribution is Gaussian, which is a distribution that can be fully described, solely by first- and second-order statistics.

for $i = 1, 2$. Plotting the signal amplitude points $(s_1(t), s_2(t))$, for $t = 1, 2, \ldots, N$ essentially produces a two-dimensional plot, also known as a scatter diagram, which in fact is a very close approximation for the true joint probability density function of the actual sources. Figure 1.1 (a) illustrates such a scatter diagram for $N = 1000$ realizations of two uniformly distributed independent random variables (s_1, s_2). Here, $s_2(t)$ is in the vertical axis and is being plotted against $s_1(t)$ represented by the horizontal axis. The joint density of the two variables closely resembles a (uniform) square, which can be easily decomposed into the product of two one-dimensional marginal densities. Since the value of each source does not convey any information about the other in this case, it is clear that the two source signals are statistically independent. After mixing the two original independent components with a random scalar mixing matrix:

$$A = \begin{pmatrix} 0.1 & 0.4 \\ 0.2 & 0.1 \end{pmatrix} \tag{1.12}$$

we can actually re-produce the set of mixtures (x_1, x_2) whose scatter plot is shown in Figure 1.1 (b). In stark contrast to Figure 1.1 (a), the marginal densities of each of the components x_1 and x_2, now appear to be mutually dependent. Although in this case, the mixed data still adheres to a uniform distribution, their joint density now takes the form of a parallelogram, which stretches accordingly depending on the specific values of the mixing matrix in Eq. (1.12). Applying PCA to the mixed data (x_1, x_2), results into two principal components. As illustrated in Figure 1.2 (a), the effect of an orthogonal transformation is a scatter plot rotation. In more detail, the scatter diagram corresponding to the whitened (uncorrelated) pair of variables (z_1, z_2), preserves the exact shape of the original source scatter plot, albeit this is now rotated by an angle with respect to the new axes. This result still differs from the original sources since the two principal components are not statistically independent. All that is left to extract the independent source components is the estimation of a single angle that provides this rotation. This can be facilitated using ICA, which not only whitens the data, but also effectively rotates it, such that the axes of the extracted source estimates (u_1, u_2) have the same direction as the original axes of (s_1, s_2). The effect of ICA is shown in Figure 1.2 (b). After applying ICA to the observed data, the recovered sources (u_1, u_2) are indeed statistically independent with a joint probability density function, which is actually identical to that of the original sources depicted in Figure 1.1 (a).

1.7   ACOUSTICS

In this booklet, we investigate one of the most commercially lucrative applications of BSS, which is the separation and recovery of a set of unknown sound sources, for example, speech signals inside an acoustically challenging and reverberant (naturally echoing) environment, using two (or more) microphones acting as sensors. In this case, each microphone (sensor) captures not only the direct contributions from each sound source but also numerous reflected and diffracted copies of the original signals, possibly at totally different propagation delays. This is due to the fact that the

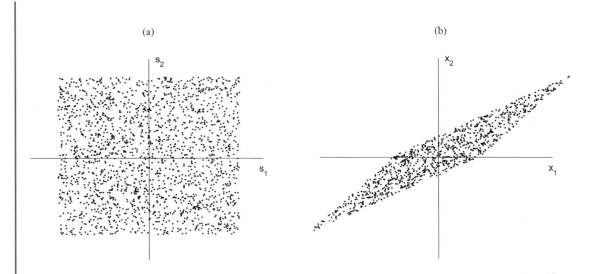

Figure 1.1: (a) The joint distribution of the independent components s_1 and s_2 with uniform distributions. (b) The joint distribution of the observed mixtures x_1 and x_2.

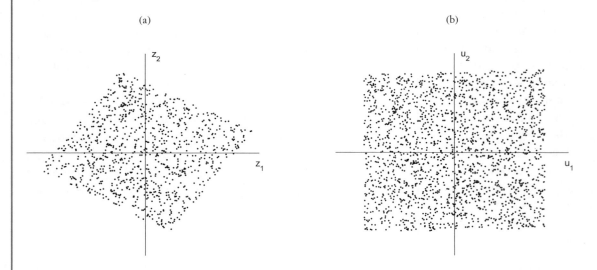

Figure 1.2: (a) The joint distribution of the uncorrelated components z_1 and z_2 after using PCA. (b) The joint distribution of the recovered (and statistically independent) source estimates u_1 and u_2 after using ICA.

original signals are corrupted from attenuation caused by loss of energy as sound propagates, as well as from multipath propagation delays arising as sound is reflected and diffracted in the medium. In effect, these delays will depend on the relative locations of the sensors and the sources and in general on the acoustic environment structure and geometry (Kuttruf, 1991).

If we assume that the signals are combined linearly, then each microphone captures the weighted sum of time-delayed versions of the original sources, which in fact is the convolution of each source with the appropriate acoustic transfer function of the room itself (Pope and Bogner, 1996b). These recorded mixtures are often referred to as the **convolutive mixtures** of the original source signals. The extraction and recovery of the individual speech sources by relying solely on the signals observed at the sensors can be accomplished through the use of BSS. In this scenario, the problem of estimating the original sources from their convolutive mixtures (observed signals), also becomes a problem of accurately modeling the acoustical properties of the enclosed space. Normally, such a task can be accomplished through the use of **finite impulse response (FIR)** filters, which appear to be ideal as candidates since (1) they provide a sufficiently accurate way to model the underlying multipath propagation characteristics for the majority of dynamic mixing conditions and (2) they are computationally simple and bounded-input bounded-output stable for bounded coefficients. In practice, the task of BSS then boils down to the estimation of these unknown transfer functions represented as FIR filters (or their inverse) by using only information contained within multiple streams of observed mixtures being recorded at the sensors. This task is carried out by blindly unraveling the effects of convolution performed by the linear time-invariant (LTI) system operating exclusively on the set of input signals (sources).

The acoustic environment plays a critical role in the separation of convolutive mixtures. Naturally, sound waves emanating from the speakers propagate through the air and are reflected and diffracted on the walls, floor and ceiling of an acoustic enclosure before they are eventually picked up by the microphone (sensor). Thus, the acoustic path itself, consists of a direct contribution and a sum of weaker and delayed contributions. In the terminology of acoustics, this is often referred to as **multipath** propagation. The acoustic path is also different for each source-to-sensor pair and in principle time-varying, unless it is assumed that the positions of the sources remain fixed. This acoustic transmission from one point in space to another can be, in general, described by its acoustic transfer function or as otherwise called **acoustic impulse response (AIR)**[7] (e.g., see Kinsler et al., 2000).

Figure 1.3 (a) depicts such an impulse response, recorded inside a normal office with the microphone placed in the center of the room, while Figure 1.3 (b) shows an impulse response captured in a very well damped room, i.e., an anechoic chamber. The impulsive excitations needed to measure the acoustic impulse responses at any given position in the room were generated using the method of frequency sweeping (Berkhout et al., 1980). In addition, both impulse responses were recorded using a sampling frequency of 8 kHz. By visually inspecting Figure 1.3 (a), the components

---

[7]This is a somewhat confusing term, since an AIR can only describe a point-to-point transfer function and it therefore insufficient to characterize a room as a whole.

of the impulse response can be divided into three groups: (1) direct sound, (2) early reflections and (3) late reflections (or reverberations). The first impulse to the microphone is the direct sound, while the second impulse which is smaller (in magnitude) is the first reflection from the surface closest to the microphone. This combined with all the rest that follow, constitute the so-called early reflections. As

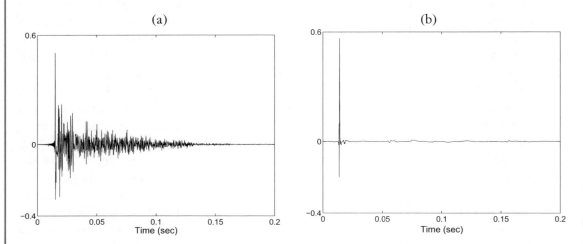

Figure 1.3: (a) Impulse response recorded inside a normal office with dimensions 4.18m × 3.80m × 2.50m and a total volume of $V = 39.71\text{m}^3$. The reverberation time T_{60} of this enclosure is approximately 100ms, a typical value for a coarsely furnitured office. (b) Impulse response recorded inside an anechoic chamber ($V = 52\text{m}^3$). Reverberation time here is almost negligible.

time elapses, the reflected sounds become smaller and smaller, since they encounter flat surfaces and get absorbed (or reflected). These impulses come closer and closer in time, meaning that the actual interval between the adjacent impulses becomes smaller and smaller until they eventually become diffused and the sound throughout the room is blended uniformly. These late reflections amount to what we perceive as reverberation (or echo) inside a room. In effect, since the number of late reflections is infinite, the tail of the room impulse response has a more dense structure.

In theory, a global characterization regarding the acoustic properties of a room can be achieved from a single, yet an important acoustic parameter, the *reverberation time*. The reverberation time (T_{60}) is defined as the interval in which the reverberating sound energy, due to decaying reflections, reaches one millionth of its initial value. In other words, it is the time it takes for the reverberation level to drop by 60 dB below the original sound energy present in the room at a given instant and is given by Kuttruf (1991):

$$T_{60} = 0.163 \, \frac{V}{\displaystyle\sum_i c_i \, A_i} \tag{1.13}$$

where $V(\mathrm{m}^3)$ is the volume of the room and $\sum_i c_i A_i$ is the effective absorption area, calculated as the sum of all surface areas $A_i(\mathrm{m}^2)$ in the room, after being multiplied by their respective absorption coefficients c_i. In effect, the acoustic character of a room is decided by its shape, layout and volume, as well as the absorption characteristics of its boundaries. In general, a rapidly decaying impulse response corresponds to a short reverberation time, while heavy tails imply longer reverberation times. Typical examples of short reverberation times include normal office rooms ($T_{60} = 100 - 200\mathrm{ms}$), whilst longer reverberation times are often associated with larger rooms, such as concert halls ($T_{60} = 0.5 - 1\mathrm{s}$) and churches ($T_{60} = 2 - 4\mathrm{s}$). For more details on advanced measurement techniques for calculating the reverberation time of an acoustical enclosure the interested reader is referred to Neubauer (2001).

In practice, the task of BSS involves estimating impulse responses which are usually several thousands of coefficients long, especially when sampled at a high sampling frequency (e.g., see Schobben *et al.*, 1999). In addition, inverting an acoustic transfer function is not always feasible. Due to obvious physical considerations, room responses are non-minimum phase, stable and causal infinite duration impulse responses, which in theory closely resemble **infinite impulse response (IIR)** filters. Hence, an accurate architecture to model the acoustic transfer functions of a reverberant room would be one that uses IIR filters. However, any filter representing a non-minimum phase system may lead to an unstable inverse system with its poles lying outside the unit circle in the z-plane (Neely and Allen, 1979). Nevertheless, when considering only the early stages of reverberation and ignoring reflections that are 60 dB below the original signal level, most environments can be approximated adequately using FIR-like responses (Miyoshi and Kaneda, 1988). Consequently, given sufficiently long filters, any room may be accurately represented using a structure of FIR filters. The advantage using such an approximation lies in the fact that FIR filters are inherently stable. In effect, it is theoretically feasible to realize an inverse stable FIR filtering solution for the non-minimum phase mixing system described by Eq. (1.24), as long as the separating filters allow for a non-causal[8] expansion.

1.8 CONVOLUTIVE MIXING MODEL

Instead of **instantaneous mixtures** as previously described in Section 1.3, we may be confronted with a set of m **observed signals** or **measurements** $\{x_i(t) \in \mathbb{R}, \ i = 1, 2, \ldots, m\}$ considered to be **linear** and **convolutive mixtures** of another set of n unknown, yet statistically independent (at each time instant) **source signals** $\{s_j(t) \in \mathbb{R}, \ j = 1, 2, \ldots, n\}$. This is often the case inside a typical acoustic environment, where each microphone (sensor) not only captures the direct contributions from each sound source but also several reflected copies of the original signals at different propagation delays. In this paradigm, the vector of signals $\mathbf{x}(t) = [x_1(t), \ldots, x_m(t)]^T \in \mathbb{R}^m$ observed at the output of

---

[8]By *non-causal*, we define a filter whose impulse response $h(t)$ is non-zero for negative time, i.e., $h(t) \neq 0$, for $t < 0$. Note that the inverse of a non-minimum phase system cannot be causal and stable at the same time.

the sensors can be modeled as:

$$x_1(t) = \sum_{k=0}^{\ell-1} h_{11}(k)\, s_1(t-k) + \ldots + \sum_{k=0}^{\ell-1} h_{1n}(k)\, s_n(t-k) \tag{1.14}$$

$$x_2(t) = \sum_{k=0}^{\ell-1} h_{21}(k)\, s_1(t-k) + \ldots + \sum_{k=0}^{\ell-1} h_{2n}(k)\, s_n(t-k) \tag{1.15}$$

$$\vdots \qquad\qquad\qquad \vdots$$

$$x_m(t) = \sum_{k=0}^{\ell-1} h_{m1}(k)\, s_1(t-k) + \ldots + \sum_{k=0}^{\ell-1} h_{mn}(k)\, s_n(t-k) \tag{1.16}$$

where t represents the discrete-time index, $\left[h_{ij}(k)\right]$ is the room impulse response, which characterizes the path between the jth source and the ith sensor for all $j = 1, 2, \ldots, n$ and $i = 1, 2, \ldots, m$ and finally $(\ell - 1)$ defines the order of the FIR filters used to model the room acoustic (or transmission channel) effects. By resorting to Eqs. (1.14)–(1.16) above, the signal $x_i(t)$ observed at the output of the ith sensor can be re-written in a more compact form as:

$$x_i(t) = \sum_{j=1}^{n} \sum_{k=0}^{\ell-1} h_{ij}(k)\, s_j(t-k), \quad i = 1, 2, \ldots, m. \tag{1.17}$$

which is valid for all $\ell > 1$. Moreover, after substituting for the filter sequence $h_{ij}(t) = \left[h_{ij}(k), h_{ij}(k-1), \ldots, h_{ij}(k-\ell+1)\right]$, the transformation imposed on the sources can essentially be seen as being equivalent to linear convolution denoted by the operator $*$:

$$x_i(t) = \sum_{j=1}^{n} h_{ij}(t) * s_j(t), \quad i = 1, 2, \ldots, m. \tag{1.18}$$

This defines the noiseless[9] linear **convolutive (or spatio-temporal) BSS mixing** model, which in matrix form is equivalent to:

$$\begin{pmatrix} x_1(t) \\ \vdots \\ x_m(t) \end{pmatrix} = \begin{pmatrix} h_{11}(t) & \cdots & h_{1n}(t) \\ \vdots & \ddots & \vdots \\ h_{m1}(t) & \cdots & h_{mn}(t) \end{pmatrix} * \begin{pmatrix} s_1(t) \\ \vdots \\ s_n(t) \end{pmatrix} \Rightarrow$$

$$\Rightarrow \quad \mathbf{x}(t) = \mathbf{H}(t) * \mathbf{s}(t) \tag{1.19}$$

_____

[9] In most applications, it would be more realistic to assume that there is some noise in the observed signal measurements, e.g., see Hyvärinen (1999) and Moreau and Pesquet (1997). In effect, this would mean adding a noise term in the BSS model shown in Eq. (1.18). Yet throughout this booklet for simplicity, we will omit any noise terms since (1) the estimation of the noiseless convolutive BSS model is difficult enough in itself and (2) it is usually impossible to separate noise from the sources without exploiting some prior knowledge, which in most cases is unavailable.

where the column vector $\mathbf{x}(t) = [x_1(t), \ldots, x_m(t)]^T$ denotes the vector of observed signals at discrete-time instant t and $\mathbf{s}(t) = [s_1(t), \ldots, s_n(t)]^T$ is the vector of the source components at the same time instant. In addition, matrix $\mathbf{H} \in \mathbb{R}^{m \times n \times \ell}$ represents the unknown linear-time in-variant (LTI) multiple-input multiple-output (MIMO) **mixing system** linking the sources with the sensor observations and is essentially a matrix of FIR filters with each element subsequently de-fined as $h_{ij}(t) = [h_{ij}(0), h_{ij}(1), \ldots, h_{ij}(\ell-1)]$, which is a row vector of length ℓ containing the mixing FIR filter coefficients. Note that this matrix notation stresses both the **multichannel** and **convolutive (or multipath)** nature of the problem.

The assumption of such a model, in which it is assumed that the transformation from the source signals to the sensors is of convolutive nature, complicates the problem considerably. Nevertheless, it does provide a sufficiently generalized (and practical) model and hence the potential to address a number of realistic acoustic scenarios, as for example multi-microphone (multi-speaker) recordings within environments which are not anechoic[10]. Figure 1.4 graphically depicts the convolutive BSS mixing and unmixing model, in the case of two-sources and two-measurements (two-microphones). By imposing certain restrictions on the (full) two-channel model, it is possible to further simplify the problem by stipulating that the direct channels can be reduced to unity transformations. This somewhat restrictive model replaces the direct filters $H_{11}(z)$ and $H_{22}(z)$ with a scaling parameter, effectively stipulating that the source signals arrive at the sensors unfiltered, which in practice cor-responds to the so-called 'separation only' problem (Lambert, 1996). Such a simplified architecture for the two-channel recording model is shown in Figure 1.5.

Alternatively, in an effort to reduce the complexity, one could easily formulate the standard convolutive BSS mixing model in the frequency-domain, instead. The premise behind such formu-lation is to make use of discrete transforms (for short-term stationary sources) in order to transform convolution operations, which are quite often performed in conjunction with long filters, into simple multiplications (Smaragdis, 1998; Parra and Spence, 2000). The z-domain equivalent of the con-volutive BSS model, effectively arises when applying the z-transform in both sides of Eq. (1.18):

$$X_i(z) = \sum_{j=1}^{n} H_{ij}(z) S_j(z), \quad i = 1, 2, \ldots, m. \tag{1.20}$$

where $X_i(z)$ and $S_j(z)$ represent the one-sided z-transforms of the discrete-time real and finite sequences $x_i(k)$ and $s_j(k)$, respectively, given by:

$$X_i(z) = \sum_{k=0}^{\infty} x_i(k) z^{-k} \tag{1.21}$$

$$S_j(z) = \sum_{k=0}^{\infty} s_j(k) z^{-k} \tag{1.22}$$

---

[10] The name 'anechoic' literally means 'without echo'. Anechoic enclosures are environments designed specifically to absorb reflections generated due to the presence of flat surfaces and are therefore built with controlled acoustic properties, in order to achieve near free-field sound propagation.

with z being a complex variable with real and imaginary parts and with z^{-k} denoting the time-shift (unit-delay) operator. In addition, $H_{ij}(z)$ defines the z-transform of the acoustic transfer function between the jth source and the ith sensor written as:

$$H_{ij}(z) = \sum_{k=0}^{\ell-1} h_{ij}(k)\, z^{-k} \tag{1.23}$$

After using the z-transform, the same model in matrix form can be re-written as:

$$\begin{pmatrix} X_1(z) \\ \vdots \\ X_m(z) \end{pmatrix} = \begin{pmatrix} H_{11}(z) & \cdots & H_{1n}(z) \\ \vdots & \ddots & \vdots \\ H_{m1}(z) & \cdots & H_{mn}(z) \end{pmatrix} \begin{pmatrix} S_1(z) \\ \vdots \\ S_n(z) \end{pmatrix} \Rightarrow$$

$$\Rightarrow \quad \mathbf{x}(z) = \mathbf{H}(z)\, \mathbf{s}(z) \tag{1.24}$$

where the column vectors of the observations and the sources present in the model, using z-transform notations, can be defined as follows:

$$\mathbf{x}(z) = [X_1(z), \ldots, X_m(z)]^T$$
$$\mathbf{s}(z) = [S_1(z), \ldots, S_n(z)]^T \tag{1.25}$$

Note also that by moving to the z-domain, the convolution operation in the model of Eq. (1.19) is now replaced with a simple multiplication[11] in Eq. (1.24).

In the most general sense, the goal of BSS in the case of convolutive mixtures is to produce a set of n signals $\{u_j(t) \in \mathbb{R},\ j = 1, 2, \ldots, n\}$, which recover and reconstruct the original unknown source signals. In practice, this is realized by finding an **unmixing** or **separating system**, the outputs of which are usually coined the **source estimates** and are described by:

$$u_j(t) = \sum_{i=1}^{m} \sum_{k=0}^{\ell-1} w_{ji}(k)\, x_i(t-k), \quad j = 1, 2, \ldots, n. \tag{1.26}$$

or, equivalently in the z-domain:

$$U_j(z) = \sum_{i=1}^{m} W_{ji}(z)\, X_i(z), \quad j = 1, 2, \ldots, n. \tag{1.27}$$

---

[11]The z-transform has a number of useful properties that follow from its definition. Yet, of particular importance when dealing with convolutive mixtures is the property dictated by the theorem of convolution, which briefly states that convolving two sequences is equivalent to multiplying their z-transforms (Oppenheim and Schafer, 1989).

In this case, $W_{ji}(z)$ defines the z-transform of the unmixing transfer function between the ith sensor and the jth source estimate written as:

$$W_{ji}(z) = \sum_{k=0}^{\ell-1} w_{ji}(k)\, z^{-k} \tag{1.28}$$

where z^{-k} denotes the time-shift (unit-delay) operator as before. Furthermore, in a similar manner to Eq. (1.24), the separating model in matrix form can be formulated as:

$$\begin{pmatrix} U_1(z) \\ \vdots \\ U_n(z) \end{pmatrix} = \begin{pmatrix} W_{11}(z) & \cdots & W_{1m}(z) \\ \vdots & \ddots & \vdots \\ W_{n1}(z) & \cdots & W_{nm}(z) \end{pmatrix} \begin{pmatrix} X_1(z) \\ \vdots \\ X_m(z) \end{pmatrix} \Rightarrow$$

$$\Rightarrow \quad \mathbf{u}(z) = \mathbf{W}(z)\,\mathbf{x}(z) \tag{1.29}$$

where the column vectors of the source estimates and the observations using z-transform notations are defined similarly to Eq. (1.25):

$$\mathbf{u}(z) = [U_1(z), \dots, U_n(z)]^T$$
$$\mathbf{x}(z) = [X_1(z), \dots, X_m(z)]^T \tag{1.30}$$

In the case of convolutive BSS, an optimal separating solution can be realized when calculating an estimate of the inverse system $\mathbf{W}(z) = \mathbf{H}^{-1}(z)$, such that the product:

$$\mathbf{G}(z) \triangleq \mathbf{W}(z)\,\mathbf{H}(z) \tag{1.31}$$

referred to as the **global system**, becomes as close as possible to an identity or unit matrix. For a separating solution, matrix $\mathbf{G}(z)$ representing the composite (cascaded) mixing-unmixing system can be approximated as $\mathbf{G}(z) \triangleq \mathbf{I}$. In this case, the recovered source estimates in the z-domain can be written as:

$$\mathbf{u}(z) = \mathbf{W}(z)\,\mathbf{H}(z)\,\mathbf{s}(z) = \mathbf{G}(z)\,\mathbf{s}(z) = \mathbf{s}(z) \tag{1.32}$$

In order to be able to guarantee the existence of $\mathbf{H}^{-1}(z)$, one must also assume that the following two conditions are satisfied (Cichocki and Amari, 2002):

C1. The mixing matrix $\mathbf{H}(z)$ is full rank on the unit circle ($|z| = 1$). This in turn implies that it is non-singular, meaning that its determinant is non-zero, such that $\det\big[\mathbf{H}(z)\big] \neq 0$ and, therefore, the matrix is invertible.

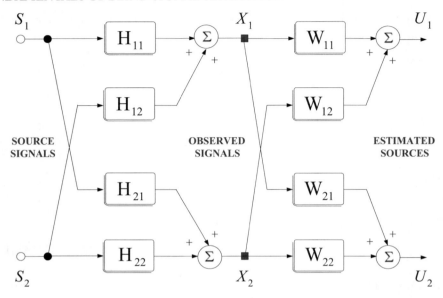

Figure 1.4: Cascaded mixing and unmixing BSS system configuration in the two-source two-sensor convolutive mixing scenario.

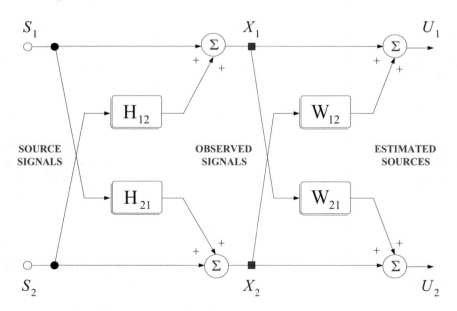

Figure 1.5: Cascaded mixing and unmixing BSS system configuration in the 'separation only' two-source two-sensor convolutive mixing scenario.

C2. Each element of the mixing matrix $\mathbf{H}(z)$ satisfies the absolute summability condition, such that:

$$\sum_{k=0}^{\ell-1} \left\| h_{ij}(k) \right\|^2 < \infty \tag{1.33}$$

which in short, states that the mixing system is stable by definition, assuming that the energy of its coefficients is finite.

In a similar manner to the decomposition of the scalar matrix $\mathbf{G}$, carried out in Eq. (1.7) (see Section 1.4) with respect to the instantaneous mixing model, the global system matrix $\mathbf{G}(z)$ in the case of convolutive BSS, can be written as:

$$\mathbf{G}(z) = \mathbf{P}\,\mathbf{\Lambda}\,\mathbf{D}(z) \tag{1.34}$$

based on which the unmixing system $\mathbf{W}(z)$ becomes equal to:

$$\mathbf{W}(z) = \mathbf{P}\,\mathbf{\Lambda}\,\mathbf{D}(z)\,\mathbf{H}^{-1}(z) \tag{1.35}$$

Note also that in Eq. (1.34) and Eq. (1.35), matrix $\mathbf{P} \in \mathbb{R}^{n \times n}$ defines the so-called permutation matrix, $\mathbf{\Lambda} \in \mathbb{R}^{n \times n}$ is a non-singular diagonal scaling matrix and finally $\mathbf{D}(z)$ represents a diagonal matrix defined as:

$$\mathbf{D}(z) = \begin{pmatrix} D_1(z) & 0 & 0 \\ 0 & \ddots & 0 \\ 0 & 0 & D_n(z) \end{pmatrix} \tag{1.36}$$

with the diagonal entries represented by a set of arbitrarily chosen filters with transfer functions subsequently given by:

$$D_i(z) = \sum_{k=0}^{\ell-1} d_i(k)z^{-k}, \quad i = 1, 2, \ldots, n \tag{1.37}$$

The indeterminacy revealed by $\mathbf{D}(z)$ implies that we can only recover a permuted (re-ordered), scaled and filtered version of the original sources from their observations. However, even when asserting that under certain conditions, the re-ordering and scaling indeterminacies can be alleviated, the indeterminacy imposed by matrix $\mathbf{D}(z)$ in the case of temporally correlated sources, e.g., speech signals, may remain unresolved. In practice, this implies that we may at best reconstruct an arbitrarily filtered version of the original sources, based on the fact that under a 'blind' setup, no additional assumptions can be made about the temporal structure of the sources beforehand.

1.9   SUMMARY

In this chapter, we have discussed the major aspects of the BSS problem, while special emphasis has been laid on the description of linear and convolutive mixtures. In fact, the process of blindly

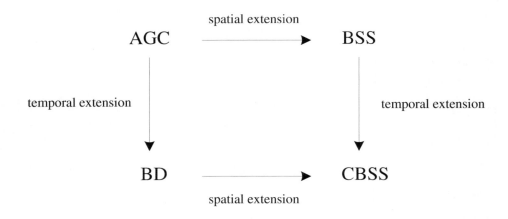

Figure 1.6: Diagram describing the relationship amongst different blind problems.

recovering the original source signals boils down to the estimation of the inverse room impulse responses. Therefore, adequately modeling the effects of the enclosed space plays a critical role. For example, in a naturally echoic setting, the mixing process takes place both temporally and spatially, which can complicate the BSS problem considerably. Below, we summarize some of the definitions concerning different aspects of the 'blind' framework. Their inherent similarities, as well as their unique relationships are also summarized in Figure 1.6 with the aid of a commutative diagram.

Convolutive blind signal separation (CBSS) is the **spatial** extension of the single-channel blind deconvolution problem where the single filter characterizing the unknown mixing system (or channel) is replaced with a matrix of filters. Alternatively, CBSS can be also viewed as the **temporal** extension of the so-called spatial-only (or instantaneous) BSS where the scalar mixing system is modeled using FIR filters arranged in a network structure. In this context, CBSS defines the **spatio-temporal** extension of BSS, in the sense that different sensors receive entirely different mixtures of the sources at different times.

Blind signal separation (BSS) is the term used to address the linear instantaneous mixing problem and describes the process of searching for a linear (and scalar) transformation that can minimize the statistical dependence between its components.

Blind deconvolution (BD) is the the single-input single-output (SISO) equivalent of CBSS, which involves blindly unraveling the effects of convolution performed by an (unknown) LTI system when operating on an input signal. In this context, BD is also referred to as blind dereverberation (e.g., see Kokkinakis and Loizou, 2009).

Automatic gain control (AGC) belongs to the degenerate case of BSS and BD where the unknown system or gain is simply described by a single scalar coefficient. The purpose here is to automatically adjust the gain of different sound sources in order to maintain a constant loudness level at the output over continuously changing conditions.

CHAPTER 2

Modern blind signal separation algorithms

2.1 INTRODUCTION

The aim of this chapter is to identify and elucidate the most significant theoretical aspects that underpin convolutive blind signal separation (CBSS). Since the assumption of statistical independence between the original sources is fundamental, it serves as the starting basis for developing different source separation strategies. We first describe the key aspects of independence, while at the same time we report on some of the most important and widely used information-theoretic BSS approaches. Due to the conceptual simplicity and intuitivity of information-theoretic criteria, we review CBSS strategies that stem from such principles. In addition, we describe CBSS strategies that rely on the use of non-stationarity and non-whiteness properties. We focus on prominent convolutive BSS approaches, which operate by solving the problem exclusively in the frequency-domain. The practical appeal of these approaches is the substantial gain in computational speed from adapting the unmixing filters in the frequency-domain. Such algorithms are considerably faster than their time domain counterparts and are, in general, more straightforward to implement.

2.2 INFORMATION-THEORETIC CRITERIA

Information theory is a useful tool for BSS, since by resorting to it one can fully characterize and measure the amount of information and redundancy existing between any two sets of signals. More precisely, starting off from an instantaneous mixing scenario, which information-theoretic principles were originally formulated for, it is argued how one can extend their validity and applicability towards solving real-world BSS problems. The scope of the following sections is twofold. It deals with the problem of choosing a suitable optimization criterion for blind signal separation, while at the same time sheds some light into some of the most important and widely used information-theoretic approaches.

2.2.1 KULLBACK-LEIBLER DIVERGENCE

A set of random variables are said to be statistically independent, if their joint distribution is a product of their marginal distributions according to Eq. (1.5)[1]. Hence, a natural way of checking whether a random variable has independent components is to simply measure the distance between

---

[1]See Section 1.4.

the two sides of Eq. (1.5) as suggested by Comon (1994). Therefore, for a vector $\mathbf{u} \in \mathbb{R}^n$ with a probability density function (PDF) denoted by $p_u(\mathbf{u})$, a widely applicable distance measure is the Kullback-Leibler divergence (KLD) (e.g., see Kullback, 1959). This statistical measure, which is otherwise known as the *relative entropy*, is essentially defined as the distance between two PDFs, $p_u(\mathbf{u})$ and $\prod_{i=1}^{n} p_{u_i}(u_i)$, such that:

$$\Delta\left(p_u(\mathbf{u}) \,\|\, \prod_{i=1}^{n} p_{u_i}(u_i)\right) = \int p_u(\mathbf{u}) \log \frac{p_u(\mathbf{u})}{\prod_{i=1}^{n} p_{u_i}(u_i)} \, d\mathbf{u} \tag{2.1}$$

In this context, KLD can be seen to reflect independence and can be considered as a relaxed concept of distance between two PDFs since it is always non-negative and zero if and only if the two distributions are equal:

$$\Delta\left(p_u(\mathbf{u}) \,\|\, \prod_{i=1}^{n} p_{u_i}(u_i)\right) \geq 0 \tag{2.2}$$

Another interesting property of the KLD is that it is invariant with respect to any invertible linear transformation, including amplitude re-scaling and permutation re-ordering. Using standard algebraic manipulations it is straightforward to show, as in Deco and Obradovic (1996), that the Kullback-Leibler divergence may be interpreted as nothing more than the average mutual information (MI) of vector $\mathbf{u}$:

$$\Delta\left(p_u(\mathbf{u}) \,\|\, \prod_{i=1}^{n} p_{u_i}(u_i)\right) = I\left(p_u(\mathbf{u})\right) = \mathrm{E}\left[\log \frac{p_u(\mathbf{u})}{\prod_{i=1}^{n} p_{u_i}(u_i)}\right] \tag{2.3}$$

2.2.2   ENTROPY MAXIMIZATION

Bell and Sejnowski (1995) were the first to frame BSS into an information-theoretic perspective. Their entropy (or information) maximization algorithm, known as INFOMAX, quickly catalyzed a surge of interest in using information theory for BSS. Initially, the main focus was to just maximize the information passed through a single neuron[2].

Yet, the authors quickly discovered that maximizing the information flow in a neural network can be interpreted as a way of approaching ICA, an observation which had been noted earlier by Nadal and Parga (1994). The crux principle here is the maximization of the mutual information $I(\mathbf{y}, \mathbf{u})$ transferred between a random input vector $\mathbf{u}$ and an invertible linear transformation of it defined as $\mathbf{y} = g(\mathbf{u}) = g(\mathbf{W}\mathbf{x})$, where $g(\cdot)$ is a nonlinear function, which approximates the (hypothesized) cumulative density function (CDF) of the unknown sources. A fairly intuitive explanation

---

[2]Neurons (or nerve cells) are a major class of cells in the nervous system, whose primary role is to process and transmit neural information. Neurons are the basic computational and information processing units of the brain and their main task is to generate and transmit electrical signals over long distances. In general, the average number of neurons in the human brain is ranging from 10^{10} to around 10^{11} (e.g., see Kandel *et al.*, 1991).

with regards to the *modus operandi* of the entropy maximization method is provided in Torkkola (1999). Let us consider a signal passing through a bounded nonlinearity g(·), which is specifically chosen to model the cumulative densities of the sources. If we now assume that the network (or neuron) outputs denoted by **u**, correspond to the source estimates, then the resulting PDF will be close to a uniform density, that is a density which exhibits the largest entropy among all bounded distributions. Therefore, maximizing the entropy of vector **y** is equivalent to recovering the original independent sources.

The effect of the nonlinear function on the probability density function of the sources is illustrated in Figure 2.1. Here, for illustration purposes the nonlinearity is represented by a logistic sigmoid function. In fact, this is equivalent to assuming a specific prior distribution for the sources, in this case, a super-Gaussian PDF, which as shown in Bell and Sejnowski (1995) works fairly well for speech signals. Consequently, the method largely operates on the implicit assumption of *a priori* knowledge regarding the (unavailable) source densities, in order to yield the PDFs of the true sources when the entropy is maximized. Essentially, this can be viewed as 'density matching' between the neuron inputs and expected outputs. Based on Eq. (2.3), the mutual information $I(\mathbf{y})$ of the network outputs can be expressed in terms of the joint and marginal differential[3] entropies as follows:

$$I(\mathbf{y}) = I(y_1, \ldots, y_n) = \sum_{i=1}^{n} H(y_i) - H(\mathbf{y}) \tag{2.4}$$

where

$$H(\mathbf{y}) = -\int p_y(\mathbf{y}) \log p_y(\mathbf{y}) d\mathbf{y} = -\mathrm{E}\big[\log p_\mathbf{y}(\mathbf{y})\big] \tag{2.5}$$

$$H(y_i) = -\int p_{y_i}(y_i) \log p_{y_i}(y_i) dy_i = -\mathrm{E}\big[\log p_{y_i}(y_i)\big] \tag{2.6}$$

In the case that operator g(·) is chosen to be a nonlinearity with a unique inverse, meaning it is either monotonically increasing or decreasing, the PDF of the network output $p_{y_i}(y_i)$ can be written as a function of the PDF of the estimated source $p_{u_i}(u_i)$ as in Papoulis (1991):

$$p_{y_i}(y_i) = \frac{p_{u_i}(u_i)}{\left|\dfrac{\partial y_i}{\partial u_i}\right|} \tag{2.7}$$

If we now combining this with Eq. (2.6) and substitute into Eq. (2.4), we can derive the following expression for the joint entropy:

$$H(\mathbf{y}) = -\sum_{i=1}^{n} \mathrm{E}\left[\log\left(\frac{p_{u_i}(u_i)}{\left|\dfrac{\partial y_i}{\partial u_i}\right|}\right)\right] - I(\mathbf{y}) \tag{2.8}$$

---

[3]The differential entropy is different from Shannon's entropy, which does not generalize for continuous distributions. Nonetheless, the differential entropy can still be interpreted as a measure of randomness, in the same way as Shannon's entropy can.

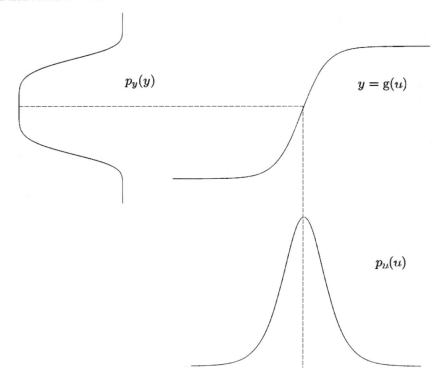

Figure 2.1: Diagram illustrating the effect of a nonlinear function on the probability density function $p_u(u)$. Top left: Uniform probability density function $p_y(y)$. Top right: The logistic function $y = g(u) = 1/(1 + e^{-u})$ approximating the cumulative density function (CDF) of the sources. Bottom right: The derivative of the logistic function $y' = g'(u) = y(1 - y)$, which closely resembles the probability density function of the original source signals denoted by $p_u(u)$.

From Eq. (2.8) it becomes apparent that a direct minimization of the mutual information $I(\mathbf{y})$ (or equivalently joint entropy maximization) can only be achieved when the marginal entropy terms are eliminated. This becomes true when the derivative of the nonlinear function $y_i = g_i(u_i)$ approximates the PDF of the estimated source $p_{u_i}(u_i)$ for all i, such that:

$$p_{u_i}(u_i) \triangleq \left| \frac{\partial y_i}{\partial u_i} \right| \qquad (2.9)$$

which clearly suggests that the overall efficiency of the optimization process closely depends on the (stipulated) density of the source estimates modeled by the nonlinear function. The strengths of the INFOMAX approach have been pointed out by both Pearlmutter and Parra (1996) and Cardoso (1997), who have independently shown that the entropy maximization approach of Bell and Sejnowski (1995) is, in fact, equivalent to the maximum likelihood (ML) estimation approach.

2.3   ADAPTIVE ALGORITHMS FOR BSS

Depending on the nature of the BSS algorithm and the application at hand, both batch and adaptive methods are commonly used in facilitating the separation stage. In an adaptive scheme, the parameters of the unmixing system are updated as each new sample (or block) is processed. This introduces another parameter, the learning rate (or step size), which affects the overall rate of convergence and must be pre-determined. In most cases, the step size is chosen empirically to be as large as possible, albeit one which still ensures that the algorithm converges. The following sections describe two of the most popular adaptive algorithms for BSS.

2.3.1   STOCHASTIC GRADIENT

Operating under the assumption that $\mathbf{u} = \mathbf{W}\mathbf{x}$ has a unique inverse[4], the multivariate probability density function can be written as in Papoulis (1991):

$$p_u(\mathbf{u}) = \frac{p_x(\mathbf{x})}{|J|} \tag{2.10}$$

where $|J|$ is the absolute value of the Jacobian matrix of the transformation. The Jacobian is the determinant of the matrix of partial derivatives given according to Bell and Sejnowski (1995) by:

$$J = \det \left[\frac{\partial u_j}{\partial x_i} \right]_{ji} \tag{2.11}$$

Further, by combining Eq. (2.5) with Eq. (2.10), the output joint entropy reduces to:

$$H(\mathbf{u}) = \mathrm{E}\big[\log |J| \big] + H(\mathbf{x}) \tag{2.12}$$

and by noting that the input entropy $H(\mathbf{x})$ remains fixed and independent of the unmixing linear transformation after invoking Eq. (2.11), Eq. (2.12) becomes:

$$H(\mathbf{u}) = \mathrm{E}\big[\log |\det(\mathbf{W})| \big] \tag{2.13}$$

Thus, from Eq. (2.4) and Eq. (2.8), it follows that the maximization of the output joint entropy $H(\mathbf{u})$ or alternatively the minimization of mutual information $I(\mathbf{u})$ can be achieved by building a simple cost (risk) log-likelihood function, such that:

$$\mathcal{G}(\mathbf{u},\, \mathbf{W}) = \sum_{i=1}^{n} \mathrm{E}\big[\log p_{u_i}(u_i) \big] + \log |\det(\mathbf{W})| \tag{2.14}$$

---

[4]See Eq. (1.6) in Chapter 1.

In this case, the gradient of the log-likelihood function becomes:

$$\frac{\partial \mathcal{G}(\mathbf{u},\ \boldsymbol{W})}{\partial \boldsymbol{W}} = \left[(\boldsymbol{W}^T)^{-1} - \varphi(\mathbf{u})\mathbf{x}^T \right] \tag{2.15}$$

To proportionally update the separating matrix $\boldsymbol{W}$ with respect to its entropy gradient in Eq. (2.15), we can resort to the **stochastic gradient** adaptation rule:

$$\boldsymbol{W}_{\ell+1} = \boldsymbol{W}_\ell + \mu \left[\boldsymbol{W}_\ell^{-T} - \varphi(\mathbf{u})\,\mathbf{x}^T \right] \tag{2.16}$$

where μ denotes a suitable step size (or learning rate) and $\varphi(\mathbf{u}) = [\varphi_1(u_1), \ldots, \varphi_n(u_n)]^T$ represents the vector of the nonlinear monotonic activation (or score) functions:

$$\varphi_i(u_i) = -\frac{\partial \log p_{u_i}(u_i)}{\partial u_i} = -\frac{\frac{\partial p_{u_i}(u_i)}{\partial u_i}}{p_{u_i}(u_i)} \tag{2.17}$$

with each element $\varphi_i(u_i)$ defined for all $i = 1, 2, \ldots, n$. For the complete derivation of the stochastic gradient method the reader is referred to Bell and Sejnowski (1995). As revealed by observing Eq. (2.9), the separation performance and convergence properties of this algorithm are in fact highly dependent upon the relation of the nonlinear activation function used in the update and its closeness to the underlying PDF of the sources to be recovered. A thorough analysis with respect to the advantages of incorporating parametric (flexible) PDF models to model the original sources is postponed until a later section.

2.3.2   NATURAL (OR RELATIVE) GRADIENT

As pointed out by Amari *et al.* (1996), the optimization parameter space, which is realized by the unmixing matrix $\boldsymbol{W}$, is not always Euclidean in nature but instead it exhibits a Riemannian metric structure. In this case, the steepest direction can be also approximated by the so-called natural gradient instead, which can be simply viewed as an optimal rescaling of the standard (stochastic) entropy gradient. Learning using the natural gradient proves to be a much more efficient option since (1) the algorithm no longer requires the computationally expensive estimation of the transpose of the inverse term $\boldsymbol{W}^{-T}$ in the update equation and (2) the steady-state performance, as well as the convergence rate are greatly improved over the standard (stochastic) gradient optimization scheme. The **natural gradient algorithm (NGA)** is derived from the standard gradient adaptation rule by post-multiplying the gradient in Eq. (2.15) with $\boldsymbol{W}^T \boldsymbol{W}$. This yields the gradient:

$$\frac{\partial \mathcal{G}(\mathbf{u},\ \boldsymbol{W})}{\partial \boldsymbol{W}} \boldsymbol{W}^T \boldsymbol{W} = \left[\boldsymbol{I} - \varphi(\mathbf{u})\,\mathbf{u}^T \right] \boldsymbol{W}_\ell \tag{2.18}$$

and subsequently the update rule:

$$\boldsymbol{W}_{\ell+1} = \boldsymbol{W}_\ell + \mu \left[\boldsymbol{I} - \varphi(\mathbf{u})\,\mathbf{u}^T \right] \boldsymbol{W}_\ell \tag{2.19}$$

where μ defines the step size (or learning rate) as before, I is the identity matrix and the column vector $\varphi(\mathbf{u}) = [\varphi_1(u_1), \ldots, \varphi_n(u_n)]^T$ represents the score functions defined in Eq. (2.17). An important milestone in the development of fast BSS algorithms was the introduction of a (slightly) modified gradient rule, the so-called **relative gradient** or as otherwise referred to the **equivariant adaptive separation via independence (EASI)** method, which was introduced independently by Cardoso and Laheld (1996). The relative gradient is characterized by a serial-updating form, whereby the decorrelation (or whitening) and the separation stages are performed simultaneously, essentially being integrated into a single-stage solution, defined as:

$$\begin{aligned} \boldsymbol{W}_{\ell+1} &= \boldsymbol{W}_\ell + \mu\, \mathcal{K}(\mathbf{u})\, \boldsymbol{W}_\ell \\ &= \boldsymbol{W}_\ell + \mu\, \left[\boldsymbol{I} - \mathbf{u}\,\mathbf{u}^T + \mathbf{u}\,\varphi(\mathbf{u})^T - \varphi(\mathbf{u})\,\mathbf{u}^T \right] \boldsymbol{W}_\ell \end{aligned} \qquad (2.20)$$

where function $\mathcal{K}(\cdot)$, depends solely on the source signal estimates and is given by:

$$\mathcal{K}(\mathbf{u}) = \boldsymbol{I} - \mathbf{u}\,\mathbf{u}^T + \mathbf{u}\,\varphi(\mathbf{u})^T - \varphi(\mathbf{u})\,\mathbf{u}^T \qquad (2.21)$$

Based on the hypotheses of (1) zero-mean and unit-variance source estimates and (2) pre-whitened mixture data, the orthogonality of the unmixing matrix is satisfied, such that $\boldsymbol{W}^{-1} = \boldsymbol{W}^T$. Thus, it is further possible to show that the relative gradient rule is in fact equivalent to the NGA update of Eq. (2.19). In this case, the following conditions are true:

$$\mathrm{E}\,[\mathbf{u}\mathbf{u}^T] = \boldsymbol{I} \qquad (2.22)$$

$$\mathrm{E}\,[\mathbf{u}\,\varphi(\mathbf{u})^T] = \boldsymbol{I} \qquad (2.23)$$

based on which, $\mathcal{K}(\cdot)$ further reduces to:

$$\mathcal{K}(\mathbf{u}) = \boldsymbol{I} - \varphi(\mathbf{u})\,\mathbf{u}^T \qquad (2.24)$$

which clearly indicates that the natural and relative gradient algorithms are in fact equivalent. Another interesting property of the aforementioned adaptive algorithms is that they belong to the class of the so-called **equivariant** algorithms. In short, this implies that their performance is completely characterized by the global system $\boldsymbol{G}$, while it is altogether independent of the individual values of the mixing matrix $\boldsymbol{A}$. To demonstrate this property, we write the NGA with respect to the global matrix:

$$\boldsymbol{G}_{\ell+1} = \boldsymbol{G}_\ell + \mu\, \left[\boldsymbol{I} - \varphi(\boldsymbol{G}_\ell\,\mathbf{s})\,\mathbf{s}^T\,\boldsymbol{G}_\ell^T \right] \boldsymbol{G}_\ell \qquad (2.25)$$

At every iteration ℓ, the expression in Eq. (2.25) depends only on the global system matrix $\boldsymbol{G}_\ell$, the source signal $\mathbf{s}$, and the step size μ, whilst it is entirely independent on the nature and equivalently the conditioning of the mixing matrix A. Nevertheless, theoretical, as well as empirical considerations, have shown that the algorithm of Bell and Sejnowski (1995) is limited to only separating sources with super-Gaussian distributions. The sigmoid function used in Bell and Sejnowski (1995) is essentially relying on *a priori* knowledge about the underlying source priors, and therefore considerably restricts the capability of the algorithm to distinguish (and hence switch accordingly) between sub- and super-Gaussian PDFs. Such a problem, can become apparent for example when sources of many different probability density functions are mixed together, in which case this type of nonlinearity is clearly inadequate.

2.4   PARAMETRIC SOURCE DENSITY MODELS FOR BSS

In recent years, much work has focused on defining mathematical models to adequately describe the amplitude distribution of a wide class of non-stationary stochastic processes, such as speech signals. Hence, lately a number of relatively well-known parametric probability density function models have started appearing in the statistical literature. In fact, early experimental studies back in the 1960s provided strong evidence that, in general, speech signals may often exhibit either a Laplacian distribution, or they, instead, resemble a generalized Gaussian form (e.g., see Davenport, 1952; Richards, 1964). In this booklet, emphasis is placed in those parametric families of densities, which can employ the aforementioned distributions, as special cases.

2.4.1   GENERALIZED GAUSSIAN DENSITY

A fairly large number of signals exhibiting unimodal and univariate densities can be approximated sufficiently by employing the **generalized Gaussian density (GGD)** model. In this context, the parametric family of densities stemming from the GGD are also known to be able to successfully model a large number of different signals, in turn comprising a fairly wide range of probability distributions. For example, in the area of image coding, GGD source modeling has been used extensively to approximate the distributions associated with the discrete cosine transform (DCT), subband coefficients of natural images, while the application of GGD has also been proven useful in video coding systems and speech modeling (e.g., see Aiazzi *et al.*, 1999; Do and Vetterli, 2002; Joshi and Fisher, 1995; Kokkinakis and Nandi, 2005; Sharifi and León-Garcia, 1995). For any zero-mean $(m_x = 0)$ signal $x \in \mathbb{R}$, the PDF of a generalized Gaussian distribution with standard deviation σ is defined as:

$$p_x(x|\nu, \sigma) = \left[\frac{\nu \cdot \mathrm{A}(\nu, \sigma)}{2 \cdot \Gamma(1/\nu)} \right] \cdot \exp\left(- \left[\mathrm{A}(\nu, \sigma) \cdot |x| \right]^\nu \right) \tag{2.26}$$

in which:

$$\mathrm{A}(\nu, \sigma) = \sigma^{-1} \left[\frac{\Gamma\left(\dfrac{3}{\nu}\right)}{\Gamma\left(\dfrac{1}{\nu}\right)} \right]^{1/2} \tag{2.27}$$

where $\sigma > 0$ and $\Gamma(\cdot)$ denotes the complete gamma function and is given by:

$$\Gamma(\lambda) = \int_0^\infty x^{\lambda-1} e^{-x} dx, \quad \lambda > 0 \tag{2.28}$$

In this case, $A(\nu, \sigma)$ is a generalized measure of the variance and essentially defines the **dispersion** or **scale** of the distribution, while the parameter ν (> 0) describes the exponential rate of decay and, in general, the **shape** of the peak at the center of the distribution $p_x(x|\nu, \sigma)$. Well-known special cases of the GGD function include a Laplacian distribution ($\nu = 1$) and a standard Gaussian or normal distribution ($\nu = 2$). In effect, smaller values of the shape parameter ν correspond to heavier tails and, therefore, to more peaked distributions. In the limiting cases, when $\nu \to +\infty$, $p_x(x|\nu, \sigma)$ converges to a uniform distribution, whereas for $\nu \to 0+$, Eq. (2.26) approaches an impulse function. The shape of the GGD for different values of the shape (or exponent) parameter ν is shown in Figure 2.2.

2.4.2   MOMENT MATCHING ESTIMATORS

For all values of $A(\nu, \sigma)$ and ν, the rth-order absolute central moment for a generalized Gaussian distributed signal is given by:

$$E[|X|^r] = \int_{-\infty}^{+\infty} |x|^r p_x(x) \, dx \tag{2.29}$$

where $E[\cdot]$ represents the expectation operator. Substituting Eq. (2.26) into Eq. (2.29), it can be shown that the rth-order statistical moments of the GGD, can be expressed with respect to the shape parameter ν of the exponent of the distribution as follows:

$$m_r = E[|X|^r] = A^{-r}(\nu, \sigma) \cdot \left[\frac{\Gamma\left(\frac{r+1}{\nu}\right)}{\Gamma\left(\frac{1}{\nu}\right)} \right] \tag{2.30}$$

which is valid for all $r > 0$. In the case of $r = 2$, the above expression produces:

$$m_2 = E[|X|^2] = \frac{1}{A^2(\nu, \sigma)} \cdot \left[\frac{\Gamma\left(\frac{3}{\nu}\right)}{\Gamma\left(\frac{1}{\nu}\right)} \right] \tag{2.31}$$

Assuming a unit-variance distribution in Eq. (2.31), the term $A(\nu)$ becomes a single variable function of ν only, yielding an expression similar to Eq. (2.27):

$$A(\nu) = \sqrt{\frac{\Gamma\left(\dfrac{3}{\nu}\right)}{\Gamma\left(\dfrac{1}{\nu}\right)}} \qquad (2.32)$$

Figure 2.3 depicts the first six moments m_r of the generalized Gaussian distribution plotted for different values of the shape parameter $\nu = 0.6, 1, 2$ and 10, in the range $r \in (1, 6)$ according to the expression derived in Eq. (2.30). The moments of the GGD are strictly monotonic and convex functions for every r and for all different distributions. Methods suggesting how to obtain a complete statistical description for distributions modeled by the GGD after matching the underlying moments of the data set with those of the assumed distribution have been recently proposed (Choi *et al.*, 2000; Kokkinakis and Nandi, 2005).

2.4.3   PARAMETRIC SCORE FUNCTIONS

The common element shared by almost all entropy-maximization based algorithms is that they remove correlations between the signals before and after applying a nonlinear function. Operating in this manner reveals the existing higher-order correlations among different signals or among different time lags of the same signal, respectively. The underlying higher-order statistics, in such case, are introduced implicitly by the use of specifically chosen nonlinear functions[5], which *parameterize* the PDF of the sources in the model. In fact, there are a number of different criteria, which dictate how to choose appropriate nonlinear functions. For instance, in the case of peaky distributions and super-Gaussian signals such as speech, theory often suggests that activation (or score) functions of a sigmoidal nature, such as the hyperbolic tangent $\varphi_i(u_i) = \tanh(u_i)$ or the standard threshold function $\varphi_i(u_i) = \text{sign}(u_i)$ are adequate choices.

On the other hand, when confronted with sub-Gaussian distributed signals, it is known that cubic nonlinearities of type $\varphi_i(u_i) = u_i |u_i|^{p-1}$ for every $p > 1$ are better candidates. Still, in most applications, the choice of an accurate score function remains a challenge, since the signal statistics are, in the majority of cases, completely unknown. The family of **parametric** nonlinear functions, based on the GGD model, can be defined by substituting for Eq. (2.26) in Eq. (2.17). First, differentiating

---

[5]Strictly speaking, the terms *nonlinear function* and *nonlinearity*, which are used interchangeably throughout this booklet, refer to $y = g(u)$, chosen to model the CDF of the sources (see Section 2.2.2). In an effort to be precise, the more accurate terms *activation* or *score function* can be used instead, which, in fact, correspond to function $\varphi(u)$ that approximates the PDFs of the source estimates, as described by Eq. (2.9) and Eq. (2.17).

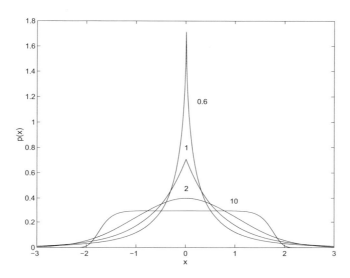

Figure 2.2: The probability density of the generalized Gaussian model plotted for different values of the shape parameter $v = 0.6, 1, 2$ and 10. All distributions are normalized to unit-variance ($\sigma^2 = 1$) and have zero-mean.

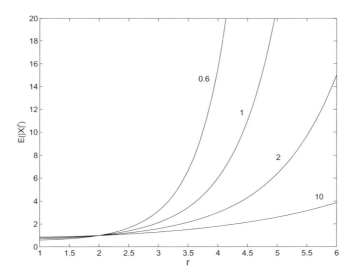

Figure 2.3: Statistical moments of the generalized Gaussian distribution plotted for different values of the shape parameter $v = 0.6, 1, 2$ and 10, in the range $r \in (1, 6)$. All distributions are normalized to unit-variance.

Eq. $(2.26)^6$ with respect to u_i, for all $i = 1, 2, \ldots, n$, yields:

$$\frac{\partial\, p_{u_i}(u_i|v_i, \sigma_i)}{\partial u_i} = v_i \cdot \left[A(v_i, \sigma_i)\, |u_i| \right]^{v_i - 1} \cdot \text{sign}(u_i) \cdot A(v_i, \sigma_i) \cdot \left[\frac{v_i \cdot A(v_i, \sigma_i)}{2 \cdot \Gamma\,(1/v_i)} \right]$$
$$\cdot \exp\left(- \left[A(v_i, \sigma_i) \cdot |u_i| \right]^{v_i} \right) \qquad (2.33)$$

Next, by resorting to the definition of $\varphi_i(u_i)$ in Eq. (2.17) and after dividing Eq. (2.33) by Eq. (2.26) and further flipping the sign, produces:

$$\varphi_i(u_i, v_i) = -\frac{\dfrac{\partial\, p_{u_i}(u_i|v_i, \sigma_i)}{\partial u_i}}{p_{u_i}(u_i|v_i, \sigma_i)} \qquad (2.34)$$

$$= v_i \cdot \left[A(v_i, \sigma_i) \cdot |u_i| \right]^{v_i - 1} \cdot \text{sign}(u_i) \cdot A(v_i, \sigma_i) \qquad (2.35)$$

$$= v_i \cdot A^{v_i}(v_i, \sigma_i) \cdot |u_i|^{v_i - 1} \cdot \text{sign}(u_i) \qquad (2.36)$$

Upon the stipulation that the sources recovered at the output have a unit-variance, such that $\sigma_i = 1$, Eq. (2.36) can be re-written as:

$$\varphi_i(u_i, v_i) = v_i \cdot A^{v_i}(v_i) \cdot |u_i|^{v_i - 1} \cdot \text{sign}(u_i) \qquad (2.37)$$

in which case, $A(v_i)$ is used instead of $A(v_i, 1)$, for simplicity. In addition, after dropping the index i and substituting for $A(v)$ using Eq. (2.32), Eq. (2.37) produces the optimal and appropriately normalized (or scaled) parametric nonlinear activation function, valid for all $v \geq 1$:

$$\varphi(u, v) = v \cdot \left[\frac{\Gamma\left(\dfrac{3}{v}\right)}{\Gamma\left(\dfrac{1}{v}\right)} \right]^{\frac{v}{2}} \cdot |u|^{v - 1} \cdot \text{sign}(u) \qquad (2.38)$$

By ignoring the scaling constant[7] above, Eq. (2.38) can be simpified even further to its so-called unscaled form, such that:

$$\varphi(u, v) = |u|^{v - 1} \cdot \text{sign}(u) \qquad (2.39)$$

In fact, when $v = 1$, Eq. (2.39) becomes equivalent to the signum function, which can also be derived from the standard Laplacian density model. For $v = 4$, $\varphi(u, v)$ yields a cubic function, which is

---

[6] See Section 2.4.1.

[7] The scaling of the recovered signals cannot be determined exactly, due to the scaling invariance, which is an inherent property of BSS. However, note that by using Eq. (2.38) instead of Eq. (2.39), it can be guaranteed that the signals recovered at the output of the separator will have a unit-variance.

suitable for sub-Gaussian distributions. Both Eq. (2.38) and Eq. (2.39) adopt a parametric structure, based entirely on the exponent parameter ν of the distribution of the source signal estimates. In principle, $\varphi(u, \nu)$ with $1 \leq \nu < 2$ can do an adequate job for super-Gaussian signals, whilst $\varphi(u, \nu)$ for $\nu > 2$ is mostly suitable for signals that exhibit a sub-Gaussian distribution. Note that in the special case when $\nu = 0.6$, essentially corresponding to a heavy-tailed (or sparse) distribution, the parametric score function $\varphi(u, \nu)$ becomes singular for $u = 0$. To circumvent such problem, Eq. (2.38) can be modified to:

$$\varphi(u, \nu) = \left[\frac{\Gamma\left(\dfrac{3}{\nu}\right)}{\Gamma\left(\dfrac{1}{\nu}\right)} \right]^{\frac{\nu}{2}} \cdot \frac{\nu\, u}{|u|^{2-\nu} + \epsilon} \tag{2.40}$$

where the definition of the signum function:

$$\mathrm{sign}(u) = \frac{u}{|u|}, \quad u \neq 0 \tag{2.41}$$

has also been taken into account. Note that in Eq. (2.40), ϵ is a very small positive parameter (typically around 10^{-4}) which, in practice, when introduced, guarantees that the singularity of $\varphi(u, \nu)$ for values around $u = 0$ is completely avoided. Based on such modification, $\varphi(u, \nu)$ remains non-singular and therefore valid, even in cases when the exponent parameter lies in the range $0 < \nu \leq 1$.

The vast majority of signals can be modeled accurately by resorting to the parametric family of generalized Gaussian densities. A characterization of statistically unknown signals can be achieved by estimating the exponent parameter of the fitted GGD model. The choice of the nonlinear activation function is a crucial factor in the success and robustness of BSS methods, which are based on entropy maximization. To this end, when parameterizing sound sources with the generalized Gaussian density model, the exponent parameter directly relates to the exponent of the score function. Prompted by this fact, we have shown that it is possible to derive optimal and flexible (parametric) nonlinearities, which can closely approximate the PDFs of the unknown sources.

2.5   BLIND SIGNAL SEPARATION STRATEGIES FOR CONVOLUTIVE MIXTURES

Broadly speaking and based upon the domain of operation, current literature on convolutive BSS can be grouped into two fairly broad classes: (1) time-domain and (2) frequency-domain approaches. Most conventional adaptive approaches for convolutive BSS attempt to perform the separation stage by operating solely in the time-domain, in which case the filter updates are carried out on a sample-by-sample basis. However, this is computationally inefficient, especially in the case of long impulse responses. The main idea behind convolutive BSS strategies operating in the frequency-domain is to make use of the short-time discrete Fourier transform (DFT) (for short-term stationary sources) in order to transform convolution operations into simple multiplications[8]. By doing so, the general problem of separating convolutive mixtures is elegantly reduced into several **independent** problems of instantaneous mixtures, one for each frequency band. Thus, a separating solution can be easily approximated with one of the many mature algorithms already proposed for the separation of instantaneous mixtures.

2.5.1   CONVOLUTIVE BSS BASED ON SECOND-ORDER STATISTICS

Non-stationarity, in the sense that source variances are time varying, is an assumption generally justified for most real-world signals, such as speech or biological signals. Since speech signals are non-stationary and uncorrelated but not (always) statistically independent, by solely relying on the assumption of non-stationarity (or alternatively non-whiteness[9]) of the underlying sources, one can focus attention towards expressing multiple cross-correlation matrices for a large number of different time intervals (e.g., see Matsuoka *et al.*, 1995; Kawamoto *et al.*, 1998). The requirement for the source estimates to be uncorrelated for all time lags is a reasonable test for independence and *second-order statistics* (SOS) constraints are sufficient for non-stationary signals to achieve BSS under some weak conditions. In fact, by simply resorting to the cross-power spectrum of the observed sensor signals at multiple times, one can build enough independent conditions to successfully *decorrelate* and hence *separate* non-stationary (e.g., speech, music) signals from their convolutive mixtures (e.g., see Parra and Spence, 2000; Schobben and Sommen, 2002; Ikram and Morgan, 2005; Pedersen *et al.*, 2007; Manmontri and Naylor, 2008).

Naturally, this leads to a joint diagonalization problem. Throughout this section, we make the following assumptions:

A1. At each time instant t, the components of the source vector $\mathbf{s}(t)$ are mutually statistically independent and therefore fulfill the condition $E\left[s_i(t)\, s_j^*(t - \ell)\right] = 0$, $i \neq j$ and $t = \ell$.

A2. The original sources are zero-mean and second-order quasi-stationary signals and their autocorrelation function is independent of a time shift, such that $\mathbf{R}_s(t) \neq \mathbf{R}_s(t + \tau)$ for some time constant $\tau \neq 0$.

A3. The number of sensors is equal to or greater than the number of the original sources, such that $m \geqslant n$.

A4. The source cross-spectral density matrix $\widetilde{\mathbf{R}}_\mathbf{s}(\omega, t)$ is diagonal for all ω and t, which follows directly from assumption A1.

---

[8]In practice, convolution operations are transformed into simple multiplications by setting the frame size T of the short-time discrete Fourier transform to be much longer than the channel filter length L, such that $T \gg L$.

[9] The *non-whiteness* assumption states that source signals are temporally correlated (or colored) signals, obeying $E\left[s_i(t)\, s_j(t - \tau)\right] \neq 0$, which is valid for $i = j$ and for some delay $\tau \neq 0$. On the other hand, the *non-stationarity* property can accommodate temporally uncorrelated (i.i.d) Gaussian sources even with identical power spectral shapes. Hence, in principle, *non-stationarity* is less restrictive than *non-whiteness*.

A5. The mixing matrix $\mathbf{A}(\omega)$ is invertible, such that $\det\left[\mathbf{A}(\omega)\right] \neq 0,\ \forall\ \omega$.

The overall objective of joint diagonalization is to find a unique matrix $\mathbf{W}(\omega)$ that simultaneously (approximately) diagonalizes a set of cross-spectral matrices over multiple observation intervals. Hence, starting from $\mathbf{u}\,(\omega, t) = \mathbf{W}(\omega)\,\mathbf{x}\,(\omega, t)$, defined for every frequency bin $\omega = 0, 2\pi/T, \ldots, 2\pi(T-1)/T$, we can obtain an estimate of the cross-spectral density matrix of the recovered sources $\widetilde{\mathbf{R}}_{\mathbf{u}}(\omega, k)$ at different super-blocks $k = 0, 1, \ldots, K-1$, written as:

$$\widetilde{\mathbf{R}}_{\mathbf{u}}(\omega, k) = \mathbf{W}(\omega)\,\widetilde{\mathbf{R}}_{\mathbf{x}}(\omega, k)\,\mathbf{W}^{H}(\omega) \tag{2.42}$$

where $(\cdot)^{H}$ denotes the Hermitian operator and $\widetilde{\mathbf{R}}_{\mathbf{x}}(\omega, k)$ denotes the cross-spectral density matrix of the observations, which can be estimated in block-form using:

$$\widetilde{\mathbf{R}}_{\mathbf{x}}(\omega, k) = \frac{1}{M} \sum_{m=0}^{M-1} \mathbf{x}(\omega, m + kM)\,\mathbf{x}^{H}(\omega, m + kM) \tag{2.43}$$

where T denotes the frame size of the DFT, $M = T(1-\beta)/(K\beta T)$ is the number of overlapping consecutive intervals used to estimate each cross-power matrix and $0 < \beta \leq 1$ is the overlapping factor. An intuitive way to calculate $\mathbf{W}(\omega)$ is to minimize a cost function that penalizes uncorrelated output signals (e.g., see Parra and Spence, 2000; Ikram and Morgan, 2005):

$$\mathcal{J}(\mathbf{W}(\omega)) = \arg\min_{\mathbf{W}(\omega)} \sum_{\omega=0}^{T-1} \sum_{k=0}^{K-1} \|\mathbf{V}(\omega, k)\|_F^2 \tag{2.44}$$

where $\|\cdot\|_F^2$ is the squared Frobenius norm (sum of squares of all elements) of the error matrix $\mathbf{V}(\omega, k) = \mathrm{off}[\widetilde{\mathbf{R}}_{\mathbf{u}}(\omega, k)]$, where operator $\mathrm{off}\,[\,\cdot\,]$ zeros the diagonal elements of matrix $\widetilde{\mathbf{R}}_{\mathbf{u}}(\omega, k)$. Based on Eq. (2.42), the term $\mathbf{V}(\omega, k)$ can be further expanded to:

$$\mathbf{V}(\omega, k) = \left[\mathbf{W}(\omega)\widetilde{\mathbf{R}}_{\mathbf{x}}(\omega, k)\mathbf{W}^{H}(\omega)\right] \\ - \mathrm{diag}\left[\mathbf{W}(\omega)\widetilde{\mathbf{R}}_{\mathbf{x}}(\omega, k)\mathbf{W}^{H}(\omega)\right] \tag{2.45}$$

where operator $\mathrm{diag}\,[\,\cdot\,]$ returns a diagonal matrix containing only the diagonal entries of the input argument[10]. In principle, we can use any suitable negative gradient search optimization technique to minimize Eq. (2.44), such as the steepest descent algorithm (Parra and Spence, 2000; Wang et al., 2005). Instead, here we choose the natural gradient method to adapt the coefficients of matrix $\mathbf{W}(\omega)$ (see Manmontri and Naylor, 2008). The most important advantages of the NGA are its simplified implementation and improved convergence accuracy and speed (see Section 2.3.2).

---

[10]Operator $\mathrm{diag}\,[\,\cdot\,]$ zeros the off-diagonal elements of matrix $\mathbf{H}$, while $\mathrm{off}\,[\,\cdot\,]$ zeros its diagonal elements, such that $\mathrm{off}\left[\mathbf{H}\right] \triangleq \mathbf{H} - \mathrm{diag}\left[\mathbf{H}\right]$.

The update equation for the joint diagonalization natural gradient algorithm (JD-NGA), after combining Eq. (2.44) with Eq. (2.45) can be expressed as follows:

$$\mathbf{W}^{\dagger}(\omega) = \mathbf{W}(\omega) + \Delta\mathbf{W}(\omega) \qquad (2.46)$$

$$\Delta\mathbf{W}(\omega) = -\gamma \, \nabla_{\mathbf{W}} \mathcal{J}(\mathbf{W}(\omega)) \qquad (2.47)$$

$$\nabla_{\mathbf{W}} \mathcal{J}(\mathbf{W}(\omega)) = 4 \sum_{k=0}^{K-1} \text{off}\left[\widetilde{\mathbf{R}}_{\mathbf{u}}(\omega, k)\right] \widetilde{\mathbf{R}}_{\mathbf{u}}(\omega, k) \, \mathbf{W}(\omega) \qquad (2.48)$$

where $\mathbf{W}^{\dagger}(\omega)$ and $\mathbf{W}(\omega)$ represent the unmixing matrix at iteration number $(\ell + 1)$ and ℓ, respectively, $\nabla_{\mathbf{W}}$ is the natural gradient operator with respect to $\mathbf{W}(\omega)$, whereas $\gamma > 0$ defines a positive constant that controls the rate of adaptation at each frequency bin ω:

$$\gamma = \frac{a}{\sum_{k=0}^{K-1} \left\| \widetilde{\mathbf{R}}_{\mathbf{x}}(\omega, k) \right\|_F^2} \qquad (2.49)$$

A summary of the JD-NGA CBSS strategy is provided in Table 2.2. Several variations of this SOS-based CBSS algorithm have been recently proposed in the BSS literature. More details can be found in Parra and Spence (2000); Joho (2000); Parra and Alvino (2002); Ikram and Morgan (2005); Wang *et al.* (2005); Manmontri and Naylor (2008).

2.5.2   CONVOLUTIVE BSS BASED ON HIGHER-ORDER STATISTICS

Convolutive BSS approaches based on *higher-order statistics* (HOS) predominantly use information-theoretic optimization criteria and therefore attempt to capture fourth-order statistics. As a result, such techniques rely heavily on the underlying statistics of the source data (e.g., see Smaragdis, 1998; Douglas and Haykin, 2000; Douglas *et al.*, 2005; Kokkinakis and Nandi, 2006).

Attempting to transform instantaneous BSS algorithms into convolutive BSS algorithms is a relatively straightforward exercise, due to the isomorphic mapping between scalar and polynomial matrices. Exploiting such an isomorphism, it can be quickly realized that different adaptation rules, such as the stochastic, natural or relative gradient updates, based on the entropy maximization principles of Bell and Sejnowski (1995) (see Section 2.2), can be easily extended to accommodate for the separation of convolutive mixtures, by applying the following three simple rules:

R1. Scalar matrices in the time-domain become equivalent to polynomial matrices in the frequency-domain.

R2. Multiplication of two scalar matrices in the instantaneous case becomes equivalent to point-wise multiplication between two polynomials in the convolutive mixing scenario.

R3. Transposition of a matrix becomes equivalent to transposition of the matrix and substitution of z by z^{-1}.

Based on these rules, the extended version of natural gradient update equation in the DFT domain, in analogy to Eq. (2.19), after dropping the z-domain operator, can be written as (e.g., see Lambert, 1996; Lambert and Bell, 1997; Lee *et al.*, 1997; Kokkinakis and Nandi, 2006):

$$\underline{\mathbf{W}}_{\ell+1} = \underline{\mathbf{W}}_{\ell} + \mu \, \Delta \underline{\mathbf{W}}_{\ell} \qquad (2.50)$$

$$\Delta \underline{\mathbf{W}}_{\ell} = \left[\underline{\mathbf{I}} - \text{FFT}\left[\varphi(\mathbf{u}) \right] \mathbf{u}^{H} \right] \underline{\mathbf{W}}_{\ell} \qquad (2.51)$$

where in this case $\underline{\mathbf{W}}$ defines the unmixing FIR polynomial matrix[11] expressed in the frequency-domain as:

$$\underline{\mathbf{W}}^{(j \times i \times k)} = \begin{pmatrix} \sum\limits_{k=0}^{\ell-1} w_{11}(k)z^{-k} & \sum\limits_{k=0}^{\ell-1} w_{12}(k)z^{-k} & \cdots & \sum\limits_{k=0}^{\ell-1} w_{1m}(k)z^{-k} \\ \sum\limits_{k=0}^{\ell-1} w_{21}(k)z^{-k} & \sum\limits_{k=0}^{\ell-1} w_{22}(k)z^{-k} & \cdots & \sum\limits_{k=0}^{\ell-1} w_{2m}(k)z^{-k} \\ \vdots & \vdots & \ddots & \vdots \\ \sum\limits_{k=0}^{\ell-1} w_{n1}(k)z^{-k} & \sum\limits_{k=0}^{\ell-1} w_{n2}(k)z^{-k} & \cdots & \sum\limits_{k=0}^{\ell-1} w_{nm}(k)z^{-k} \end{pmatrix} \qquad (2.52)$$

with each FIR polynomial given by[12]:

$$W_{ji}(z) = \sum_{k=0}^{\ell-1} w_{ji}(k)z^{-k}, \qquad (2.53)$$

defined for all $j = 1, 2, \ldots, n$ and $i = 1, 2, \ldots, m$, which in essence, describes a moving-average (MA) process. In addition, the operators FFT[$\cdot$] and IFFT[$\cdot$] represent the Fast Fourier and the Inverse Fast Fourier Transform, respectively, defined here as element-wise operations on the filter, whereas the term FFT [$\varphi(\mathbf{u})$] denotes the frequency-domain representation of the nonlinear activation (score) function $\varphi(\mathbf{u})$, such that:

---

[11]The term FIR polynomial matrix is used here to denote either the z-domain or the short-time discrete Fourier domain representation of any scalar matrix in the frequency-domain. For more details on this type of representation, which has since been adopted by many researchers working in the area of convolutive BSS, the reader is referred to Lambert (1996).

[12]See also Eq. (1.28) defined in Section 1.8.

Table 2.1: Stochastic, natural and relative gradient updates

Algorithm	Update equation in the frequency-domain
INFOMAX	$\underline{\mathbf{W}}_{\ell+1} = \underline{\mathbf{W}}_{\ell} + \mu \left[\underline{\mathbf{W}}_{\ell}^{-H} - \text{FFT}\left[\varphi(\mathbf{u})\right] \mathbf{x}^{H} \right]$
NGA	$\underline{\mathbf{W}}_{\ell+1} = \underline{\mathbf{W}}_{\ell} + \mu \left[\mathbf{I} - \text{FFT}\left[\varphi(\mathbf{u})\right] \mathbf{u}^{H} \right] \underline{\mathbf{W}}_{\ell}$
EASI	$\underline{\mathbf{W}}_{\ell+1} = \underline{\mathbf{W}}_{\ell} + \mu \left[\mathbf{I} - \mathbf{u}\mathbf{u}^{H} + \mathbf{u}\,\text{FFT}\left[\varphi(\mathbf{u})\right]^{H} - \text{FFT}\left[\varphi(\mathbf{u})\right] \mathbf{u}^{H} \right] \underline{\mathbf{W}}_{\ell}$

$$\text{FFT}\left[\varphi(\mathbf{u})\right] = \left(\text{FFT}\left[\varphi_1(u_1)\right], \dots, \text{FFT}\left[\varphi_n(u_n)\right] \right)^{T} \tag{2.54}$$

where in turn $\varphi(\mathbf{u})$ operates in the time-domain and is defined as in Eq. (2.17) or as in Eq. (2.38) with respect to its parametric (generalized) form. In addition, $(\cdot)^{H}$ is the Hermitian (or complex conjugate transpose) operator and the identity (unit) FIR polynomial matrix symbolized by $\mathbf{I}$ is given by:

$$\mathbf{I} = \begin{bmatrix} \bar{1} & \bar{0} \\ \bar{0} & \bar{1} \end{bmatrix} \tag{2.55}$$

where $\bar{1}$ and $\bar{0}$ represent a sequence of all ones and all zeros, respectively. The source estimates represented by vector $\mathbf{u}(z)$ are defined similarly to Eq. (1.30). In essence, the update rules in Eqs. (2.50)–(2.51) facilitate the frequency-domain representation of the natural gradient algorithm (FD-NGA) for CBSS. For more details about variations of this HOS-based CBSS algorithm, the reader is referred to Douglas and Haykin (2000); Grbić *et al.* (2001); Douglas *et al.* (2005); Kokkinakis and Nandi (2006); Kokkinakis and Loizou (2007); Lee *et al.* (2008).

Other update rules, which correspond to *extended* versions of the update rules previously described for instantaneous mixtures[13] are summarized in Table 2.2. A summary of the FD-NGA CBSS strategy is also provided in Table 2.3. Such methods exhibit significant advantages over conventional time-domain-based CBSS techniques, in that even fairly long unmixing filters, which are very common in speech separation tasks, can be most efficiently adapted in the frequency-domain using standard and well-established algorithms. The most important advantages of re-formulating the CBSS problem in the frequency-domain are the following:

- The capability of reducing filter convolutions to simple element-wise multiplications, in order to yield fast operations in the frequency-domain.

---

[13] See Section 2.3.

- The ability to define inverse non-causal[14] filters for non-minimum mixing phase systems, assuming that enough zeros have been pre-pended and post-pended to allow for an accurate expansion.

2.5.3   SCALING AND PERMUTATION

One can see that operating in the frequency-domain, exhibits significant advantages. Nonetheless, attempting to perform convolutive BSS in the frequency-domain does not come without perils of its own. A major concern documented for the aforementioned methods is that they all tend to suffer from **scaling** and **permutation** indeterminacies. Both are inherent, yet trivial ambiguities in the standard instantaneous mixing model (see Section 1.4). However, when performing CBSS independently in each frequency band, both ambiguities are transformed into major inconsistencies which affect separation performance. More analytically:

Scaling. The variances (energies) of the recovered sources in each frequency band cannot be uniquely determined. The variant scaling among different frequency bins can result in spectral deformation of the original sounds during the recovery process since the reconstructed signals are, in fact, arbitrarily filtered versions of the original sources.

Permutation. The order of the recovered sources in each frequency band cannot be identified exactly. These disparities can appear due to misaligned permutations produced along the frequency axis making the time-domain reconstruction of the recovered sources problematic. Hence, signals will remain mixed unless correctly aligned.

Removing both scaling and permutation indeterminacies is crucial since the sources in the time-domain are essentially obtained from recovered outputs in all the frequencies. Over the last few years, a number of approaches have addressed ways to cope with the scaling and permutation indeterminacies. In more detail:

Solutions to scaling. Applying post-filtering to the separated signals in the time-domain (e.g., see Pedersen *et al.*, 2007), normalizing the separating filters to a unit-norm in the frequency-domain (e.g., see Smaragdis, 1998; Schobben and Sommen, 2002), constraining the diagonal elements of the separating filters to unity (e.g., see Parra and Spence, 2000), enforcing the so-called minimal distortion principle (e.g., see Matsuoka and Nakashima, 2001), or even inverse filtering using the separating filters just after the separation stage separately in each frequency band (e.g., see Ikeda *et al.*, 2001) are some of the many solutions proposed to solve the scaling indeterminacy issue.

[14]See Section 1.7 for a definition.

Solutions to permutation. Spectral smoothing by averaging separation matrices with adjacent frequencies (e.g., see Smaragdis, 1998), imposing constraints on the separating filter lengths in the time-domain (e.g., see Parra and Spence, 2000; Schobben and Sommen, 2002), utilizing multi-stage beamforming techniques and direction of arrival (DOA) estimation (e.g., see Parra and Alvino, 2002; Ikram and Morgan, 2005), the combination of beamforming with subspace methods (for equal number of sources and sensors) (e.g., see Mitianoudis and Davies, 2003), alongside with inter-frequency correlation of the signal envelopes (e.g., see Araki *et al.*, 2007) (for even more than two sources) and even a hybrid of the two aforementioned approaches (e.g., see Pedersen *et al.*, 2007) are just some of the ideas proposed to mitigate permutation misalignment effects.

2.6   PERFORMANCE MEASURES

In general, when attempting to assess the separation quality in quantitative terms, it quickly becomes obvious that there is no single perfect measure of goodness since, in fact, depending on the desired objective, there are more than one definitions for the problem itself. For example, in the case of instantaneous mixtures, in most cases, it suffices to just observe the global (or equivalently the scaled permutation matrix) and attempt to quantify performance by simply measuring their 'distance' from the (desired) identity matrix. On the other hand, when attempting to evaluate the efficiency of BSS algorithms in the case of convolutive mixtures, we have to cope with a certain level of complexity often present in the real-world, which complicates the evaluation process considerably.

Since, in most practical applications, the channel impulse responses and source signals are not known, one can assess the performance of the algorithm with the **signal-to-interference-ratio improvement (SIRI)** by measuring the overall amount of crosstalk reduction achieved by the CBSS algorithm before and after the unmixing (or separation) stage, which in dB is equal to:

$$\text{SIRI} = 10 \log \left(\sum_{i=1}^{m} \frac{\sum_{\omega=0}^{T-1} |u_{ii}(\omega)|^2}{\sum_{\substack{j=1 \\ j \neq i}}^{n} \sum_{\omega=0}^{T-1} |u_{ij}(\omega)|^2} \right) - 10 \log \left(\sum_{i=1}^{m} \frac{\sum_{\omega=0}^{T-1} |x_{ii}(\omega)|^2}{\sum_{\substack{j=1 \\ j \neq i}}^{n} \sum_{\omega=0}^{T-1} |x_{ij}(\omega)|^2} \right) \tag{2.71}$$

where $x_{11}, x_{22}, u_{11}, u_{22}$ represent the direct-channel and $x_{12}, x_{21}, u_{12}, u_{21}$ the cross-channel individual contributions of the original sources, realized by creating two set of mixtures and source estimates after assuming that only one of the sources becomes active at each one time. The SIRI is particularly well-suited for speech signals, since the signatures of the original sources are reconstructed back to each microphone and therefore remain invariant to any filtering indeterminacies that could be imposed by the CBSS strategy.

Step 1. Initialize the separating filters in the frequency-domain, such that $\mathbf{W}^{(0)}(\omega) = \mathbf{I}$.

Step 2. For every super-block $k = 1, 2, \ldots, K - 1$:

Compute the DFT of the ith mixed signal observed at the sensor input:

$$x_i(\omega, t) = \sum_{\tau=0}^{T-1} w(\tau) x_i(t + \tau) e^{-J\, 2\pi\omega\tau} \tag{2.56}$$

where $w(\tau)$ represents a Hanning sliding window defined as:

$$w(\tau) = 0.5 \left[1 - \cos\left(2\pi\tau/T - 1\right)\right], \quad \forall\ \tau = 0, 1, \ldots, T - 1. \tag{2.57}$$

Step 3. Estimate the cross-spectral density matrix according to:

$$\widetilde{\mathbf{R}}_{\mathbf{u}}(\omega, k) = \mathbf{W}(\omega)\, \widetilde{\mathbf{R}}_{\mathbf{x}}(\omega, k)\, \mathbf{W}^H(\omega) \tag{2.58}$$

Step 4. Build the cost function:

$$\mathbf{V}(\omega, k) = \mathrm{off}\left[\mathbf{W}(\omega)\widetilde{\mathbf{R}}_{\mathbf{x}}(\omega, k)\mathbf{W}^H(\omega)\right] \tag{2.59}$$

Step 5. Build the correction function:

$$\mathbf{C}(\omega) = 4 \sum_{k=0}^{K-1} \mathrm{off}\left[\widetilde{\mathbf{R}}_{\mathbf{u}}(\omega, k)\right] \odot \widetilde{\mathbf{R}}_{\mathbf{u}}(\omega, k) \odot \mathbf{W}(\omega) \tag{2.60}$$

where $\odot$ denotes element-by-element multiplication between vectors.

Step 6. Compute the gradient in the frequency-domain and update the elements of the separating filters, such that:

$$\Delta\mathbf{W}(\omega) := -\gamma\,\mathbf{C}(\omega) \tag{2.61}$$

where γ denotes the chosen step size parameter.

Step 7. Estimate the jth source signal from the mixtures in the frequency-domain:

$$u_j(\omega, t) = \sum_{i=1}^{m} W_{ji}(\omega) \odot x_i(\omega, t), \quad j = 1, 2, \ldots, n. \tag{2.62}$$

Step 8. Output the kth block of the jth recovered source signal estimate in the time-domain:

$$u_j(t) = \frac{1}{T} \sum_{\omega=0}^{T-1} u_j(\omega, t)\, e^{J\, 2\pi\omega\tau} \tag{2.63}$$

Step 9. Return to Step 2 and increment the super-block number. Repeat above until convergence.

Table 2.2: Summary of the joint diagonalization natural gradient CBSS algorithm

Step 1. Initialize the separating filters in the frequency-domain.

Step 2. For every block $k = 1, 2, \ldots, L - 1$:

Compute the DFT of the ith mixed signal observed at the sensor input:

$$X_i(k) = \text{FFT}\left[\underbrace{x_i(k-1)L \ldots x_i(kL-1)}_{(k-1)\text{-th block}} \quad \underbrace{x_i(kL) \ldots x_i(kL+L-1)}_{k\text{-th block}} \right]^T \qquad (2.64)$$

Step 3. Estimate the jth source signal from the mixtures in the frequency-domain, which according to (1.27) is equal to:

$$U_j(k) = \sum_{i=1}^{m} W_{ji}(k) \odot X_i(k), \quad j = 1, 2, \ldots, n. \qquad (2.65)$$

where $\odot$ denotes element-by-element multiplication between the two vectors.

Step 4. Calculate the kth block of the jth source estimate in the time-domain to drive the nonlinearity:

$$u_j(k) = \text{last } L \text{ terms of IFFT}\left[U_j(k) \right] \qquad (2.66)$$

Step 5. Choose a suitable score function to approximate the PDF of the source, for example the parametric function described in Eq. (2.38) or the hypertangent function, which is equal to:

$$\varphi_i\left[u_i(k) \right] = \tanh(\cdot) \qquad (2.67)$$

Step 6. Formulate the (temporary) frequency-domain vector of the nonlinearity $\varphi_i(\cdot)$, such that:

$$\Phi_i(k) = \text{FFT}\left[\underbrace{0 \cdots 0}_{L \text{ zeros}} \quad \varphi_i\left[u_i(kL) \right] \quad \underbrace{\varphi_i\left[u_i(kL) \right] \ldots \varphi_i\left[u_i(kL+L-1) \right]}_{k\text{-th block}} \right]^T \qquad (2.68)$$

Step 7. Compute the gradient in the frequency-domain and update the elements of the separating filters, such that:

$$\Delta W_{ji}(k) := \mu\left[W_{ji}(k) - \Phi_i(k) \odot U_j^*(k) \odot W_{ji}(k) \right] \qquad (2.69)$$

where $(\cdot)^*$ denotes complex conjugation and μ represents a pre-determined step size parameter.

Step 8. Output the kth block of the jth recovered source signal estimate in the time-domain:

$$u_j(k) = \text{first } L \text{ terms of IFFT}\left[U_j(k) \right] \qquad (2.70)$$

Step 9. Return to Step 2 and increment the block number. Repeat above until convergence.

Table 2.3: Summary of the frequency-domain natural gradient CBSS algorithm

Although the SIRI metric can measure channel identification accuracy reasonably well, it might not always reflect the output speech quality. A disadvantage of the SIRI is that it simply cannot provide any information regarding the overall distortion present in the enhanced signal. For these reasons, we may also assess the performance of any CBSS algorithm using the **perceptual evaluation of speech quality (PESQ)** (ITU-T, 2001). The PESQ employs a sensory model to compare the original (unprocessed) with the enhanced (processed) signal, which is the output of the CBSS algorithm, by relying on a perceptual model of the human auditory system. More precisely, the PESQ can be calculated as a linear combination of the average disturbance value D_{ind} and the average asymmetrical disturbance value A_{ind}:

$$\text{PESQ} = a_0 + a_1 \, D_{\text{ind}} + a_2 \, A_{\text{ind}} \tag{2.72}$$

such that $a_0 = 4.5$, $a_1 = -0.1$ and $a_2 = -0.0309$. By definition, a high value of PESQ indicates low speech signal distortion, whereas a low value suggests high distortion with considerable degradation present. On average, PESQ values normally range between 1.0 and 4.5, thus resembling a five grade mean opinion score (MOS) scale.

2.7   SUMMARY

In recent years, numerous algorithms have been proposed for blind separation of convolutive mixtures. In general, most BSS algorithms exhibit a satisfactory performance when the mixing process is stationary and instantaneous, but only a few strategies are capable of operating in real-world and non-optimal listening scenarios. In this chapter, we have presented and extensively discussed two of the most important CBSS strategies. More precisely, we have focused on frequency-domain algorithms, which are the current state-of-the-art for real-time implementations. In the next chapter, we focus on the application of such modern CBSS strategies to noise reduction for the hearing-impaired.

CHAPTER 3

Application of blind signal processing strategies to noise reduction for the hearing-impaired

3.1 INTRODUCTION

This chapter focuses on multi-microphone blind signal processing strategies suitable for implementation in behind-the-ear (BTE) processors of hearing aid (HA) and cochlear implant (CI) devices. Most modern BTE processors come furnished with two or more microphones and with all the electronic components integrated in a housing behind the ear, and, therefore, they offer the capacity to integrate intelligent multi-microphone noise reduction strategies in order to enhance possibly noisy incoming signals. For such devices, we show how BSS can be used to improve speech understanding and potentially increase listening comfort for hearing-impaired individuals and specifically HA and CI users when communicating in noisy and reverberant environments. Furthermore, we discuss in great detail the most significant theoretical and practical aspects that underpin multi-microphone processing strategies for noise reduction in HAs and CIs. In comparison to single-microphone noise reduction strategies, which typically use only spectral and temporal information, multi-microphone processing strategies can make use of spatial information due to the relative position of the sound sources. Such approaches generally improve the signal-to-noise ratio (SNR) performance, especially when the speech and noise sources are spatially separated.

3.1.1 HEARING LOSS

In general, depending on which part of the ear is affected, hearing loss can be classified into *conductive* and *sensorineural* (Moore, 2007). Conductive hearing loss is caused by problems in the outer and middle ear interfering with the transmission of sound to the inner ear. This type of hearing loss is the most common cause of hearing impairment especially in children. The *conductive* component of the hearing loss describes the blockage of sounds from reaching the sensory cells of the inner ear. In conductive hearing loss, the inner ear functions normally, but sound vibrations are blocked from passage through the ear canal, ear drum or across the tiny bones located in the middle ear. Conductive hearing loss is usually mild to moderate in degree, may occur in one or both ears at the same time, and, in most cases, is correctable by relatively minor medical and surgical treatments or simple by mere amplification of sound (e.g., see Palmer and Ortmann, 2005; Kates, 2008).

Sensorineural hearing loss[1] is commonly referred to as nerve deafness or sensorineural deafness and is the result of disorders in the inner ear, the cochlea or the auditory nerve pathways from the inner ear to the brain. About 90% of all reported hearing loss cases belong to this category. Sensorineural hearing loss is mainly caused by damage to hair cells in the cochlea and essentially renders the conversion of sound from mechanical movement to neural activity unfeasible. Due to the fact that cochlear hair cells, once destroyed, do not regenerate, this type of hearing loss is permanent. The most common cause of sensorineural hearing impairment is a progressive age-related hearing loss called presbycusis (Lipkin and Williams, 1986). This type of hearing loss is caused by the gradual loss and subsequent death of the delicate hair cells within the inner ear. Other common causes for the destruction of hair cells are congenital defects, excessive exposure to noise and certain viral infections or side-effects from intravenous medication.

The damage to the hair cells associated with sensorineural hearing loss produces the following changes in the perception of sound (1) increased hearing threshold or, in other words, loss of sensitivity to weak sounds, (2) reduced frequency selectivity which contributes mainly to difficulty in discriminating the spectral shape of speech and (3) reduced dynamic range. As described by Moore (2003), depending on the amount of inner and outer hair cell damage, hearing loss varies from mild (20-40 dB), moderate (40-70 dB), severe (70-90 dB) to profound (> 90 dB).

3.1.2   HEARING AIDS

Mild to severe hearing loss can be partially compensated for by the use of hearing aids (HAs). Hearing aids are the single most effective therapeutic approach for the majority of people with hearing loss. Hearing aids are ear-level or body-worn instruments designed to amplify sound (Kates, 2008). Hearing aids record sounds in the acoustical environment through one or more microphones, then amplify those sounds and following amplification they direct the amplified signal into the ear canal of the listener through a loudspeaker, which is also known as the receiver. Hearing aids work differently depending on the electronics used.

The two main types of electronics are analog and digital. Almost all HAs currently available today are digital hearing aids (Palmer and Ortmann, 2005). These hearing aids contain a programmable chip which can analyze incoming sound and convert the sound to a digital signal. The signal is then manipulated based on the characteristics of the incoming sound and each individual's hearing levels and listening needs. HAs that process sound digitally offer the potential of considerable improvements over previously available analog instruments and many users report a subjective preference for digital hearing aids due to the sensation of a more natural and comfortable sound. In

---

[1]Sensorineural hearing loss is one of the most prevalent chronic health conditions affecting approximately 32 million Americans every year. Almost one in five individuals is affected with some form of hearing loss by age 55. In the long term, hearing loss can have enormous social and economic impacts, limiting education, employment and social participation of otherwise healthy individuals (e.g., see Jackson and Turnbull, 2004).

behind-the-ear digital hearing aids, all electronic components are integrated in a housing that can be placed behind the ear.

3.1.3 COCHLEAR IMPLANTS

Hearing aid devices have improved significantly over the years. Nonetheless, they still operate on the premise that the inner hair cell function remains intact. Therefore, individuals suffering from profound hearing loss who have little inner and outer hair cells left usually receive no benefit with a hearing aid. In this case, if the auditory nerve is still intact, cochlear implants (CIs) can be used instead to partially restore hearing sensation (e.g., see Loeb, 1990; Loizou, 1998). The cochlear implant[2] is a small complex electronic device that is implanted under the scalp with electrodes positioned in the cochlea to stimulate the ganglion cells of the auditory nerve (Clark, 2003). Electrical current induces action potentials in the auditory nerve fibers and these are then transmitted directly to the auditory cortex where they are interpreted as sound. Thus, CIs can bypass damaged or missing hair cells within the cochlea that would normally code sound.

Cochlear implants have both internal and external components. The external components include a microphone and a speech processor placed behind the outer ear. An additional part of CIs is an external transmitter that is attached to the outside of the skin. Directly opposite the external transmitter is an internal receiver (cochlear stimulator) placed beneath the scalp and implanted into the bone behind the ear. Starting from the internal receiver that is implanted in the mastoid bone an array of stimulating electrodes wraps partly around the scala tympani of the cochlea. The incoming signals picked up by the speech processor are transmitted to the internal coil, and then the appropriate electrode along the array is selected and stimulated at a level that most closely approximates the loudness of the incoming signal.

Cochlear implants have progressed from being just a speculative laboratory procedure to a widely recognized practice, a proven medical option, and a life-changing use of technology. Nowadays, cochlear implant devices provide sound awareness to a vast majority of profoundly deaf individuals (e.g., see House and Berliner, 1991; Gates and Miyamoto, 2003). In recent years, it has become common practice to place cochlear implants in both sides of an individual patient either simultaneously (same surgery) or sequentially (separate surgeries at different times). As in normal-hearing (NH) individuals, binaural hearing is far superior to monaural hearing and offers considerable advantages over unilateral implantation (e.g., see van Hoesel and Tyler, 2003; Litovsky *et al.*, 2004; Litovsky *et al.*, 2009). Bilateral CI recipients tend to receive significant benefit, such that most adult users understand the majority of speech and can communicate with minimal or no lip reading, especially in quiet listening environments. Today, bilateral cochlear implants (BI-CIs) are offered to

---

[2]According to the Food and Drug Administration (FDA), as of April 2009, approximately 190,000 people worldwide have received implants. In the U.S., roughly 45,000 adults and around 25,500 children have received cochlear implants. There are currently three CI systems in use in the U.S. (1) the Clarion cochlear implant manufactured by the Advanced Bionics Corporation in Sylmar, CA (2) the Nucleus 24 cochlear implant developed by the University of Melbourne, Australia and the Cooperative Research Center (CRC) for Cochlear Implant Speech and Hearing Aid Innovation manufactured by Cochlear Corporation and (3) the Combi 40+ system manufactured by MED-EL in Austria.

a growing number of individuals, including adults and children, in order to provide benefits arising from having two functional ears instead of one[3]. Bilateral CI users can nowadays achieve word recognition scores of almost 80% or even higher regardless of the device used (e.g., see Spahr and Dorman, 2004).

3.2 SPEECH INTELLIGIBILITY IN NOISE

Hearing-impaired (HI) listeners have greater difficulty understanding speech in noise than do people with normal hearing abilities. In fact, poor speech intelligibility in noisy environments is a major source of dissatisfaction for HI individuals. A standard way to quantify speech intelligibility in noise is the so-called *speech reception threshold* (SRT), which is the signal-to-noise ratio required to achieve a 50% intelligibility score. High SRT values correspond to poor speech intelligibility (Smits *et al.*, 2006). In practice, HI listeners need approximately 3-6 dB higher SNR than normal-hearing (NH) listeners in order to perform at the same level in typical noisy backgrounds (Moore, 2007). For speech-shaped noise (SSN), which is a steady noise with a long-term spectrum matching that of natural speech, the SRT increase for HI listeners has been found to be ranging from 2.5 dB to 7 dB, while for fluctuating noise or competing speech, the increase is considerably higher (Plomp, 1994). In the presence of a single talker, the increase in SRT for HI listeners can be as much as 15 dB (Peters *et al.*, 1998).

A number of studies reveal a similar trend in both unilateral and bilateral CI recipients (e.g., see Spahr and Dorman, 2004; Stickney *et al.*, 2004; Loizou *et al.*, 2009; Kokkinakis and Loizou, 2008, 2010). Overall, speech recognition in quiet is considerably better than speech recognition even in moderate noise. Spahr and Dorman (2004) have shown that for speech material presented at 10 dB SNR, the average speech intelligibility performance of CI recipients can decrease to 70% on tasks using clean speech and to around 40% during tasks involving conversational speech. At 5 dB SNR, recognition of conversational speech sentences fall on average to around 20%. Since, for typical speech materials, just a 1 dB increase in SRT leads to a 10-20% reduction in the overall percent correct score from the above data it quickly becomes evident that even a 2-3 dB elevation in SRT can create significant problems when attempting to understand speech in background noise.

To improve speech intelligibility in noisy conditions, a number of single microphone noise reduction techniques have been proposed over the years (e.g., see Hochberg *et al.*, 1992; Weiss, 1993; Loizou *et al.*, 2005; Yang and Fu, 2005; Loizou, 2006; Hu *et al.*, 2007). While single microphone noise reduction techniques can achieve only moderate improvements in speech intelligibility, considerably larger benefits can be obtained when resorting to multi-microphone adaptive signal processing strategies, instead. Such strategies make use of spatial information due to the relative position of the emanating sounds and can therefore better exploit situations in which the target and masker are spatially (or physically) separated. Nowadays, most BI-CIs are fitted with either two microphones in each ear or one microphone in each of the two (one per ear) BTE processors.

---

[3]According to the databases of the three major cochlear implant manufacturers, there are approximately 3,000 bilaterally implanted individuals worldwide as of October 2005 and 58% of these are children.

The Nucleus Freedom processor, for instance, employs a rear omni-directional microphone, which is equally sensitive to sounds from all directions, as well as an extra directional microphone pointing forward[4]. Resorting to directional microphones provides an effective yet simple form of spatial processing. A number of recent studies have shown that the overall improvement provided by the use of an additional directional microphone in the device can be 3-5 dB in real-world environments with relatively low reverberant characteristics when compared to processing with just an omni-directional microphone (Wouters and van den Berghe, 2001; Chung et al., 2006).

Adaptive beamformers can be considered an extension of differential microphone arrays, where the suppression of interferers is carried out by adaptive filtering of the microphone signals. An attractive realization of adaptive beamformers is the generalized sidelobe canceller (GSC) structure (Griffiths and Jim, 1982). To evaluate the benefit of noise reduction for CI users, van Hoesel and Clark (1995) tested a two-microphone noise reduction technique, based on adaptive beamforming by mounting a single directional microphone behind each ear. The results obtained indicated large improvements in speech intelligibility for all CI subjects tested, when compared to an alternative two-microphone strategy, in which the inputs to the two microphone signals were simply added together.

The performance of beamforming algorithms in various everyday-life noise conditions using BI-CI users was also assessed by Hamacher et al. (1997). The mean benefit obtained in terms of the speech reception threshold with the beamforming algorithms for four BI-CI users varied between 6.1 dB for meeting room conditions to just 1.1 dB for cafeteria noise. In another study, Chung et al. (2006) conducted experiments to investigate whether incorporating directional microphones and adaptive multi-channel noise reduction algorithms could enhance overall CI performance. The results indicated that directional microphones can provide an average improvement of around 3.5 dB. An additional improvement of approximately 2 dB was observed when processing the noisy stimuli through the front directional microphone first and then through the noise reduction algorithm.

In another study, Wouters and van den Berghe (2001) assessed speech recognition of four adult CI users utilizing a two-microphone adaptive filtering beamformer. Monosyllabic words and numbers were presented at 0° azimuth at 55, 60, and 65 dB SPL in quiet and noise with the beamformer inactive and active. Speech-shaped noise was presented at a constant level of 60 dB SPL from a source located at 90° azimuth on the right side of the user. Word recognition in noise was significantly better for all presentation levels with the beamformer active, showing an average SNR improvement of more than 10 dB across conditions. More recently, Spriet et al. (2007) investigated the performance of the BEAM pre-processing strategy in the Nucleus Freedom speech processor with five CI users. The performance with the BEAM strategy was evaluated at two noise levels and with two types of noise, speech-shaped noise and multi-talker babble. On average, the algorithm

---

[4]Microphones operate by sensing the pressure difference on either side of a thin sheet known as a diaphragm. Ultimately, there are really only two fundamental microphone types — *omni-directional* and *directional*. Omni-directional microphones are designed to be equally sensitive to sounds arriving from all directions, without essentially aiming to favor one direction over another, while directional microphones are sensitive to sounds emanating from only one direction and reject sounds coming from different azimuths outside the desired pickup area (polar pattern) (e.g., see Chung, 2004).

tested, lowered the SRT by approximately 5-8 dB, as opposed to just using a single directional microphone to increase the direction-dependent gain of the target source.

It is clear from the above studies that processing strategies based on beamforming can yield substantial benefits in speech intelligibility for cochlear implant users, especially in situations where the target and masker sound sources are spatially separated. Nevertheless, the effectiveness of beamforming strategies is limited to: (1) only zero-to-moderate reverberation settings (e.g., see Greenberg and Zurek, 1992; Hamacher *et al.*, 1997; Kompis and Dillier, 1994) and (2) only a single interfering source. In fact, both the overall reverberation time of the room and the direct-to-reverberant ratio (DRR) have been found to have the strongest impact on the performance of adaptive beamformers (Kompis and Dillier, 1994). Also, the presence of multiple noise sources can considerably reduce the overall efficiency and performance of beamforming algorithms. A substantial drop in SRT performance in the presence of uncorrelated noise sources originating from various azimuths has been noted by both Spriet *et al.* (2007) and van den Bogaert *et al.* (2009).

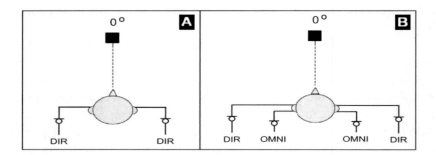

Figure 3.1: (a) Two-microphone (one per side) bilateral and (b) four-microphone (two per side) bilateral configuration.

The majority of the aforementioned studies evaluated noise reduction strategies in situations where a single interfering source was present and the room acoustics were characterized by low or no reverberation (anechoic settings). In realistic scenarios, however, rooms might have moderate to high reverberation and multiple noise sources might be present, and in some instances, these sources might be emanating from both hemifields (left and right). In the following sections, we report on adaptive multi-microphone noise reduction strategies, which (1) can address realistic scenarios and (2) can be implemented in the latest commercial hearing aid and cochlear implant devices. Special emphasis is placed on a multi-sensor array configuration, whereby speech is assumed to be collected simultaneously over several (two or more) spatially distributed microphones located in each of the two (one per ear) BTE processors worn by the hearing-impaired user. Stimuli processed in single and multi-source anechoic and reverberant scenarios are used in our evaluation.

3.3   NOISE REDUCTION STRATEGIES FOR HEARING-IMPAIRED LISTENERS

A significant part of this chapter is dedicated on multi-microphone noise reduction strategies for bilateral cochlear implant devices. Note that similar noise reduction strategies can be also easily applied to bilateral hearing aids. As depicted in Figure 3.1, the following two configurations are investigated, based on the total number of microphones available:

Two-microphone bilateral. In this configuration, we assume access to a total of two directional microphones (left and right) with each microphone placed on opposite sides of the head (Kokkinakis and Loizou, 2008).

Four-microphone bilateral. In this configuration, we assume access to a total of four microphones, namely two omni-directional (left and right) and two directional microphones (left and right) with each set of directional and omni-directional microphones placed on opposite sides of the head (Kokkinakis and Loizou, 2010).

3.3.1   2M-SESS STRATEGY

The two-microphone spatial enhancement via source separation (2M-SESS) strategy can be classified as a two-microphone binaural strategy since it relies on having access to a total of two directional microphones (left and right) with each microphone placed on opposite sides of the head. This configuration is illustrated in Figure 3.1 (a). As outlined in the diagram shown in Figure 3.2, the 2M-SESS strategy operates by estimating a total of four adaptive linear filters that can undo the mixing effect by which two composite signals are created when the target and noise sources propagate inside a natural acoustic environment. The 2M-SESS strategy can spatially separate and further suppress noise, and hence the CI user can maximize speech intelligibility by focusing only on the extracted target source. Although in principle, one can allow the CI user to select which enhanced signal output to listen to, the 2M-SESS strategy is designed to reject the interferer and deliver only the recovered speech diotically to the user.

To adaptively estimate the unmixing filters, the 2M-SESS strategy employs the frequency-domain natural gradient algorithm (FD-NGA)[5] (e.g., see Kokkinakis and Loizou, 2008). The 2M-SESS strategy runs in an adaptive *off-line* mode and relies on a multi-pass processing scheme. Therefore, the filter estimation is performed iteratively over a block of data and the estimates obtained in the last iteration are then used to process the same data blocks. That is, the same blocks of data are re-used. This is a strategy typically employed in adaptive algorithms to obtain sufficiently good estimates of the filters. An *off-line* scheme is not amenable to real-time implementation, but it is used here to gauge on the upper performance bound assuming that large amounts of training data are available.

---

[5] See Chapter 2, Section 2.5 and Table 2.3 for a detailed description of this algorithm.

3.3.2   2M2-BEAM STRATEGY

The 2M2-BEAM noise reduction strategy utilizes the two-microphone BEAM running in an independent stimulation mode. As shown in Figure 3.1 (b), the 2M2-BEAM employs two (one per ear) BTE units furnished with one directional and one omni-directional microphone and is therefore capable of delivering the signals bilaterally to the CI user. Each BEAM combines a directional microphone with an extra omni-directional microphone placed closed together in an endfire array configuration to form the target and noise references. The inter-microphone distance is usually fixed at 18 mm (Patrick *et al.*, 2006).

In the BEAM strategy[6], the first stage utilizes spatial pre-processing through a single-channel, adaptive dual-microphone system that combines the front directional microphone and a rear omni-directional microphone to separate speech from noise. The output from the rear omni-directional microphone is filtered through a fixed finite impulse response filter. The output of the FIR filter is then subtracted from an electronically delayed version of the output from the front directional microphone to create the noise reference (Wouters and van den Berghe, 2001; Spriet *et al.*, 2007). The filtered signal from the omni-directional microphone is then added to the delayed signal from the directional microphone to create the speech reference. This spatial preprocessing increases sensitivity to sounds arriving from the front while suppressing sounds that arrive from the sides. The two signals with the speech and noise reference are then fed to an adaptive filter, which is updated with the normalized least-mean-squares (NLMS) algorithm (Haykin, 1996) in such a way as to minimize the power of the output error (Greenberg and Zurek, 1992). The 2M2-BEAM strategy is currently implemented in commercial bilateral CI processors, such as the Nucleus 24 Freedom processor.

3.3.3   4M-SESS STRATEGY

The 4M-SESS processing strategy is the two-stage multi-microphone extension of the conventional BEAM noise reduction approach. It is based on co-ordinated or cross-side stimulation by collectively using information available on both the left and right sides. The 4M-SESS strategy can be classified as a four-microphone binaural strategy since it relies on having access to four microphones, namely two omni-directional (left and right) and two directional microphones (left and right) with each set of directional and omni-directional microphones placed on opposite sides of the head. This arrangement, depicted in Figure 3.1 (b), is already available commercially in the Nucleus 24 Freedom implant processor.

As outlined schematically in Figure 3.3, in the 4M-SESS binaural processing strategy, the speech reference on the right side is formed by adding the input to the left omni-directional microphone to the delayed version of the right directional microphone signal and the noise reference on the right is estimated by subtracting the left omni-directional microphone signal from a delayed version of the right directional microphone signal. In a similar manner, to create the speech reference signal on the left side, the signals from the left directional microphone right and omni-directional microphone are summed together. The noise reference on the left side is formed by subtracting the

---

[6]For a more thorough explanation of the BEAM strategy, the reader is referred to Spriet *et al.* (2007).

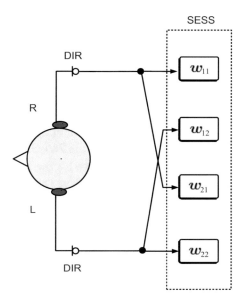

Figure 3.2: Block diagram of the 2M-SESS adaptive noise reduction processing strategy.

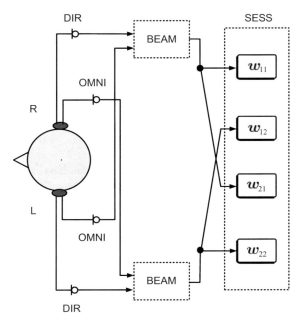

Figure 3.3: Block diagram of the 4M-SESS adaptive noise reduction processing strategy.

right omni-directional microphone signal from a delayed version of the left directional microphone signal. Assuming that the noise source is placed on the right of the listener, this procedure leads to a signal with an amplified noise level on the right side but also yields an output with a substantially reduced noise level in the left ear. After processing the microphone signals containing two speech and two noise reference signals binaurally with one BEAM processor per ear, the two microphone outputs from the two BEAM processors containing the generated speech and noise reference signals are enhanced further by employing the joint diagonalization natural gradient algorithm (JD-NGA)[7].

The 4M-SESS strategy is the four-microphone on-line extension of the two-microphone algorithm in Kokkinakis and Loizou (2008) and unlike the 2M-SESS strategy is amenable to real-time implementation. The 4M-SESS strategy operates by estimating a total of four FIR filters that can undo the mixing effect by which two composite signals are generated when the target and noise sources propagate inside an acoustic environment. The filters are computed after only a single pass with no additional training. The 4M-SESS strategy operates on the premise that the target and noise source signatures are spatially separated, and thus their individual form can be retrieved by minimizing the statistical dependence between them. In statistical signal processing theory, this configuration is referred to as a fully determined system where the number of independent sound sources is equal to the number of microphones available for processing. Initializing the unmixing filters used in the 4M-SESS strategy with those obtained with each of the two BEAM pre-processing blocks results in a substantial reduction in the total number of filter coefficients required for adequate interference rejection and speeds up the convergence of the algorithm. Similarly to the 2M-SESS strategy, the implementation of the 4M-SESS strategy requires access to a single processor driving two CIs, such that signals from the left and right sides are captured synchronously and processed together.

3.4   SPEECH INTELLIGIBILITY STUDIES WITH HEARING-IMPAIRED LISTENERS

3.4.1   SUBJECTS

The subjects participating in our studies were all native American English speaking adults with post-lingual deafness who received no benefit from hearing aids pre-operatively. All subjects were cochlear implant listeners fitted bilaterally with the Nucleus 24 multi-channel implant device manufactured by Cochlear Corporation. The participants used their devices routinely and had a minimum of 4 years experience with their CIs. Biographical data for the subjects tested is provided in Table 3.1.

All subjects tested were using the Cochlear Esprit BTE processor on a daily basis. During their visit, the participants were temporarily fitted with the SPEAR3 wearable research processor. SPEAR3 was developed by the Cooperative Research Center (CRC) for Cochlear Implant and Hearing Aid Innovation, Melbourne, Australia, in collaboration with HearWorks. The SPEAR3 has been used in a number of investigations to date as a way of controlling inputs to the cochlear

---

[7]See Chapter 2, Section 2.5 and Table 2.2 for a detailed description of this algorithm.

Table 3.1: Cochlear implant patient description and history

	S1	S2	S3	S4	S5
Age	40	58	36	68	70
Gender	F	F	F	M	F
Years of CI experience (L / R)	6 / 6	4 / 4	5 / 5	7 / 7	8 / 8
Years of deafness	12	10	15	11	22
Etiology of hearing loss	Unknown	Unknown	Noise	Rubella	Hereditary

implant system (e.g., see van Hoesel and Tyler, 2003). Prior to the scheduled visit of the subjects, the Seed-Speak GUI application was used to program the SPEAR3 processor with the individual threshold (T) and comfortable loudness levels (C) for each participant.

3.4.2 PROCEDURE

All cochlear implant listeners tested used the device programmed with the advanced combination encoder (ACE) speech coding strategy (Vandali et al., 2000). In addition, all parameters used (e.g., stimulation rate, number of maxima, frequency allocation table, etc.) were matched to each patient's clinical settings. The volume of the speech processor was also adjusted to a comfortable loudness prior to initial testing. Before participants were enrolled in this study institutional review board approval was obtained and before testing commenced informed consent was obtained from all participants.

Each participant completed testing in eight sessions of 2.5 hours each spanning several days. At the start of each session, all participants were given a short practice session in order to gain familiarity with the task. Separate practice sessions were used for different conditions. No score was calculated for these practice sets. The subjects were told that they would hear sentences in a noisy background, and they were instructed to type what they heard via a computer keyboard. It was explained that some of the utterances would be hard to understand and that they should make their best guess.

To minimize any order effects during testing, such as learning or fatigue effects, all conditions were randomized among subjects. Different sets of sentences were used in each condition. After each test session was completed, the responses of each individual were collected, stored in a written sentence transcript and scored off-line by the percentage of the keywords correctly identified. All words were scored. The percentage correct speech recognition scores were calculated by dividing the number of key words the listener repeated correctly by the total number of key words in the particular sentence list.

3.4.3   STIMULI

The speech stimuli used for testing were sentences from the IEEE database (IEEE, 1969). Each sentence is composed of approximately 7 to 12 words, and in total, there are 72 lists of 10 sentences each produced by a single talker. The root-mean-square amplitude of all sentences was equalized to the same root-mean-square value, which corresponds to approximately 65 dBA. Every sentence in the IEEE speech corpus that was produced by a male talker was designated as the target speech.

In order to simulate the speech interferer or competing voice in this experiment, a female talker uttering the sentence *"Tea served from the brown jag is tasty"* (also taken from the IEEE database) was chosen as the female masker (or non-target). The other interferer used in our experiments was speech-shaped noise generated by approximating the average long-term spectrum of the speech to that of an adult male taken from the IEEE corpus. Both target and masker speech had the same onset, and where deemed necessary, the interferers were edited to have equal duration to the target speech tokens. All the stimuli were recorded at a sampling frequency of 16 kHz.

3.4.4   ANECHOIC CONDITIONS

A set of free-field-to-eardrum (or anechoic) head-related transfer functions (HRTFs) measured in an acoustic manikin (Head Acoustics, HMS II.3), as described in the AUDIS catalogue (see Blauert *et al.*, 1998), were used to simulate different spatial locations of the speech target and the masker signals. These recordings were carried out in an anechoic chamber designed to absorb most of the sound reflections that are normally present. The length of the HRTFs was 256 sample points, amounting to a relatively short delay of 16 ms and no reverberation.

To generate the multi-sensor composite (or mixed) signals observed at the pair of microphones, the target and masker stimulus for each position were *convolved* with the set of HRTFs for the left and right ear, respectively. For this particular experiment, the target speech source was assumed to be placed directly in front of the subject at 0° azimuth at the realistic conversational distance of 1 m. To generate stimuli in various spatial configurations, we set the azimuth angles of the masker positions to the right of the listener at 0°, 30°, 45°, 60° and 90°. In all cases, the vertical position of the sources was adjusted at 0° elevation. A diagram illustrating this configuration is provided in Figure 3.4.

To evaluate speech intelligibility in these anechoic settings, the following conditions were used for each masker type and masker azimuth angle: (1) binaural unprocessed and presented bilaterally and (2) diotic stimulation plus noise reduction using the 2M-SESS processing strategy. Hence, in total there were 20 different conditions (2 maskers × 5 spatial configurations × 2 strategies) using a total of 40 sentence lists. In the binaural unprocessed case, the two simulated sensor observations captured from one microphone were fed to one ear and, similarly, the composite signals observed in the other microphone, were presented to the other ear via the auxiliary input jack of the SPEAR3 processor. In the processed case, the enhanced signal was presented diotically to the bilateral users via the auxiliary input jack of the SPEAR3 processor.

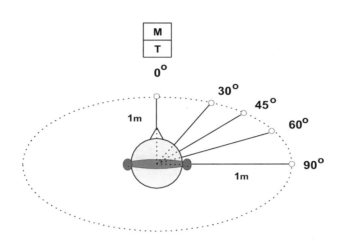

Figure 3.4: Setup of the spatial configuration used in the experiments with target (T) at 0° azimuth and with each of the maskers (M) arranged in a semicircle of 1 m radius at different azimuth angles to the right of the listener.

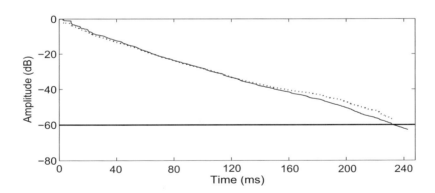

Figure 3.5: Energy decay curves of the left-ear (solid line) and right-ear (dash line) impulse responses at 60°. The time taken for the amplitude to drop by 60 dB (thick line) below the original sound energy level is around 200 ms.

3.4.5 REVERBERANT CONDITIONS

To simulate more realistic and challenging reverberant conditions a different set of HRTFs was measured using a CORTEX MKII manikin artificial head inside a mildly reverberant room with dimensions $5.50 \times 4.50 \times 3.10 \, \mathrm{m}^3$ (length × width × height). The room was fitted with acoustical

Table 3.2: List of spatial configurations tested

Number of interferers	Type of interferer	Left or distributed on both sides	Right or distributed on right
One interferer	Speech-shaped noise		90°
Three interferers	Speech-shaped noise	-30°, 60°, 90°	30°, 60°, 90°

curtains to change its acoustical properties (e.g., see van den Bogaert *et al.*, 2009). The HRTFs were measured using identical microphones to those used in modern BTE speech processors. HRTFs provide a measure of the acoustic transfer function between a point in space and the eardrum of the listener and also include the high-frequency shadowing component due to the presence of the head and the torso. The length of the impulse responses was 4,096 sample points at 16 kHz sampling rate amounting to an overall duration of approximately 0.25 s. The reverberation time of the room was experimentally determined following the ISO 3382 standard by using an omni-directional point source B&K 4295 and an omni-directional microphone B&K 2642. As illustrated in Figure 3.5, the reverberation time was found to be approximately equal to 200 ms when measured in octave bands with center frequencies of 125-4000 Hz.

The artificial head was placed in the middle of a ring of 1.2 m inner diameter. Thirteen single-cone loudspeakers (FOSTEX 6301B) with a 10 cm diameter were placed every 15° in the frontal plane. A two-channel sound card (VX POCKET 440 DIGIGRAM) and DIRAC 3.1 software type 7841 (Bruel and Kjaer Sound and Vibration Measurement Systems) were used to determine the impulse response for the left and right ear by transmitting a logarithmic frequency sweep. In order to generate the stimuli recorded at the pair of microphones for each angle of incidence the target and interferer stimuli were convolved with the set of HRTFs measured for the left- and right-hand ear, respectively. In total, there were four different sets of impulse responses for each CI configuration employing two-microphones. In a similar manner as before, all stimuli were presented to the listener through the auxiliary input jack of the SPEAR3 processor in a double-walled sound attenuated booth (Acoustic Systems, Inc). During the practice session, the subjects were allowed to adjust the volume to reach a comfortable level during delivery in both ears.

The simulated target location was always placed directly in front of the listener at 0° azimuth. Subjects were tested in conditions with either one or three interferers. In the single interferer conditions, a single speech-shaped noise source was presented from the right side of the listener (+90°). In the conditions where multiple interferers were present, three interfering noise sources were placed asymmetrically either across both hemifields (−30°, 60°, 90°) or distributed on the right side only

(30°, 60°, 90°). Table 3.2 summarizes these experimental conditions. In the single interferer condition, the initial SNR was set at 0 dB, whereas for multiple interferers, the initial SNR for each interferer was fixed at 5 dB, and hence the overall level of the interferers was naturally increased as more interferers were added.

The noisy stimuli in the case of a single interferer were processed with the following stimulation strategies: (1) unilateral presentation using the unprocessed input to the directional microphone on the side ipsilateral to the noise source, (2) bilateral stimulation using the unprocessed inputs from the two directional microphones, (3) bilateral stimulation plus noise reduction using the 2M2-BEAM strategy and (4) diotic stimulation plus noise reduction using the 4M-SESS processing strategy. The noisy stimuli generated when three interferers were originating from either the right or from both hemifields were processed with the following processing strategies: (1) bilateral stimulation using the unprocessed inputs from the two directional microphones, (2) bilateral stimulation plus noise reduction using the 2M2-BEAM strategy and (3) diotic stimulation plus noise reduction using the 4M-SESS processing strategy. In total, there were 8 different conditions (4 strategies × 2 signal-to-noise-ratios) for the single interferer scenario and 6 different conditions (3 strategies × 2 spatial configurations) tested with multiple interferers. A total of 28 IEEE sentence lists were used. Two IEEE lists (20 sentences) were used for each condition.

3.5   RESULTS & DISCUSSION

3.5.1   ANECHOIC CONDITIONS

Speech intelligibility scores obtained in the anechoic conditions (see Section 3.4.4) are shown in Figure 3.6. The scores obtained using the unprocessed sentences were higher in the 90° condition, where the masker and target signals were spatially separated, than in the 0° condition, in which case the masker and target signals originated from the same location. This suggests that all bilateral cochlear implant listeners tested were able to benefit from spatial release of masking[8]. This observation is consistent with previous studies on the same subject (e.g., see Tyler et al., 2002; van Hoesel and Tyler, 2003; Stickney et al., 2004). The observed spatial unmasking, however, seems to be largely dependent on the spatial separation between the masker and target signals, as expected (Hawley et al., 2004). In stark contrast, sentences processed by the 2M-SESS processing strategy were only mildly dependent on the physical separation between the masker and target signals.

Based on the results plotted in Figure 3.6, we can also conclude that as long as the separation between the target and masker signals is at least 30° or more, the 2M-SESS processing strategy can produce large improvements in intelligibility. In the noise masker condition, for instance, word recognition scores improved from roughly 40% correct with unprocessed sentences to 80% correct when the corrupted stimuli was first processed with the 2M-SESS strategy. Large improvements in

---

[8]The benefit in perceived speech intelligibility introduced by the spatial separation between the speech and noise sources has been termed as spatial release from masking or spatial unmasking (e.g., see Saberi et al., 1991; Hawley et al., 2004). This improvement can be measured by computing the difference in speech intelligibility between spatially separated conditions and spatially co-located conditions.

performance were obtained with the 2M-SESS strategy for both types of interferers (female taker and speech-shaped noise). Spatially segregating the target speech from its respective maskers by filtering the corrupted composite signals through a set of FIR unmixing filters results in a compelling release from masking. From a theoretical standpoint, the fact that the 2M-SESS strategy performs equally well in both informational and energetic masking[9] conditions and for all almost spatial configurations is to be anticipated, since the underlying 'blind' algorithm utilizes no prior knowledge with regard to the original signals or their underlying mixing structure.

3.5.2 EFFECTS OF TRAINING ON SPEECH INTELLIGIBILITY

Nonetheless, given that the 2M-SESS strategy requires no previous information on the specifics of the acoustical setup, some amount of training is essential in order to achieve a considerable amount of masker suppression. In the present experiment, a total of 60 sec was required to achieve an adequate level of performance. Logically, this raises the question of the amount of training required for the 2M-SESS strategy to produce a reasonable separation performance and further to achieve similar speech intelligibility scores as the ones previously obtained (see Figure 3.6).

   To thoroughly investigate the effect of training on speech intelligibility for bilateral CI users, the same strategy was re-applied to enhance the male target embedded in female speech and speech-shaped noise, synthesized binaurally with HRTFs. The same subjects were used and an identical procedure to the one described in Section 3.4 was followed. The main difference here was that training was not carried out individually for every single speech token as before. Instead, the unmixing filters were adapted just for a randomly selected set of signals. After convergence to a separating solution, the unmixing filters were saved, and then without any further modification used to enhance the remaining sentences.

   Note that, in fact, we employed the *same* set of estimated filters to enhance signals embedded either in female or noise. The rationale behind this approach is that the 2M-SESS strategy should ideally remain truly "blind" to the original sources, and hence performance should not suffer. Based on this strategy, only a limited number of filters, namely one set for every spatial position of the maskers is required. This results in considerable savings in processing time. To further assess to what degree training affects separation quality the algorithm is allowed only 2, 3 and 5 passes (or iterations) through the data, which in effect correspond to 5, 10, and 15 sec of total training time.

   The results obtained for different training times are given in Figure 3.7 for just two spatial configurations (30° and 90°). The data obtained in Figure 3.6 with 60 sec of training are also included for comparative purposes. The results indicate that in three out of all four conditions tested, the performance obtained after 10 sec of training was not significantly different to the performance

---

[9]The presence of any acoustic source, other than the target signal provides an energy floor masking the source. If the level of the signal is too close to this energy floor, the peripheral neural activity elicited by the competing noise will overwhelm that of the target signal. Thus, it is difficult to detect and comprehend a speech signal when other sound sources (speech or non-speech) are present. This sort of masking is referred to as *energetic* (e.g., see Freyman *et al.*, 2001). A speech masker normally interferes with the perception and recognition of the targeted speech at both peripheral and cognitive levels. In the literature, any central level interference resulting from stimulus (speech or non-speech sound) uncertainty is referred to as *informational* masking (e.g., see Arbogast *et al.*, 2002).

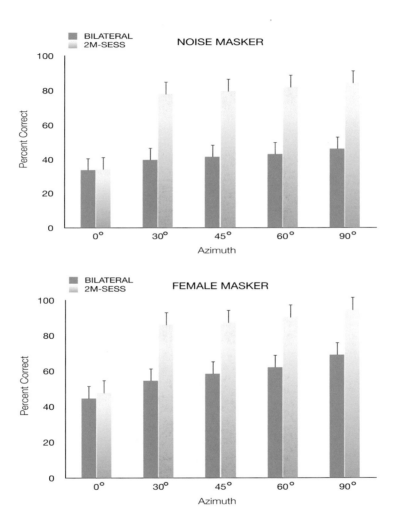

Figure 3.6:   Mean percent word recognition scores for five Nucleus 24 CI users tested on IEEE sentences embedded in speech-shaped noise (top) and female speech (bottom) at SNR = 0 dB. Scores for sentences presented using bilateral stimulation and the default processor strategy are shown in blue. Scores for sentences processed first through the 2M-SESS strategy, and then the default processor strategy and presented diotically to the CI users are shown in yellow. Error bars indicate standard deviations.

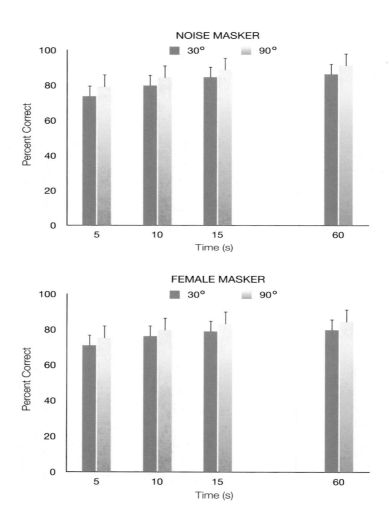

Figure 3.7: Mean percent word recognition scores plotted against training (or adaptation) time for five Nucleus 24 CI users on IEEE sentences embedded in speech-shaped noise (top) and female speech (bottom) at SNR = 0 dB. Scores for sentences processed first through the 2M-SESS processing strategy, and then the default processor strategy for a masker source placed at 30° azimuth are shown in green. Scores for sentences processed first through the 2M-SESS processing strategy, and then the default processor strategy for a masker source placed at 90° azimuth are plotted in yellow. Error bars indicate standard deviations.

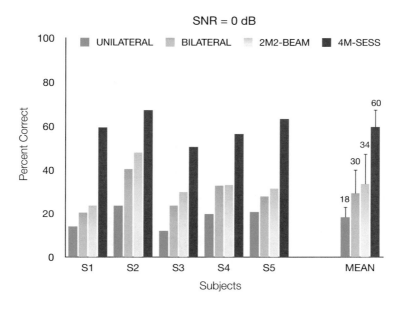

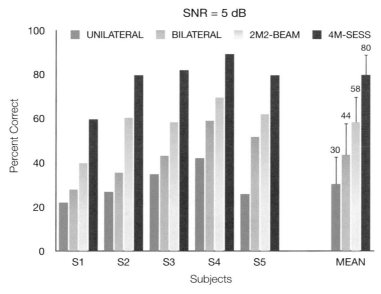

Figure 3.8: Percent correct scores by five Nucleus 24 users using both CI devices tested on IEEE sentences originating from the front of the listener (0° azimuth) and embedded in a single speech-shaped noise source placed on the right (+90°). Error bars indicate standard deviations.

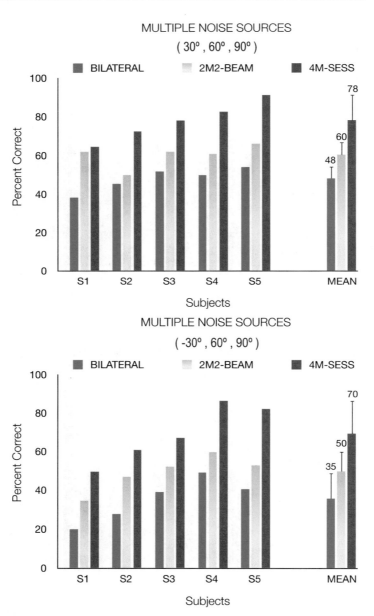

Figure 3.9: Percent correct scores by five Nucleus 24 users using both CI devices tested on IEEE sentences originating from the front of the listener (0° azimuth) and embedded in three speech-shaped noise sources distributed on the right side only (30°, 60°, 90°) or placed asymmetrically across both the left and right hemifield (−30°, 60°, 90°). Error bars indicate standard deviations.

obtained after a total of 60 sec of training. This finding implies that 10 sec of training seems to be sufficient in as far as reaching an asymptotic performance, at least in most conditions. The only exception to this hypothesis was observed for the female masker condition at 90° azimuth, which required 15 sec of training to reach a performance level equal to that obtained with 60 sec of overall data training.

From Figure 3.7, we can draw the following conclusions. First, as expected from theory, by increasing the adaptation time and hence the available signal length, the separation performance improves. This is reflected by the high recognition scores obtained when a total of 60 sec of training is performed. Second, the 2M-SESS processing strategy required no more than a few seconds of data (5-10 sec) in order to converge to a solution yielding an audibly distinguishable performance. Such observation can be confirmed by Figure 3.7, which shows relatively high word recognition scores for the 10 sec case, for both types of maskers and for both the 30° and 90° azimuths.

3.5.3   REVERBERANT CONDITIONS

Speech intelligibility scores obtained in simulated reverberant conditions (see Section 3.4.5) with the 2M2-BEAM and 4M-SESS multi-microphone adaptive noise reduction strategies are shown in Figure 3.8 (one interferer) and Figure 3.9 (three interferers). As shown in Figure 3.8, when the single noise source arrives from the side ipsilateral to the unilateral cochlear implant, speech understanding in noise was quite low for all subjects when presented with the input to the fixed directional microphone. Although one would expect that processing with a directional microphone alone would result in a more substantial noise suppression this is not the case here. The small benefit observed is probably due to the high amount of reverberant energy present since most directional microphones are designed to only work efficiently in relatively anechoic scenarios or with a moderate levels of reverberation (e.g., see Hawkins and Yacullo, 1984). In contrast, improvements in speech understanding were obtained when BI-CI listeners used both cochlear implant devices. When compared to the unprocessed unilateral condition, the mean improvement observed was 12 percentage points at 0 dB SNR and 14 percentage points due to processing when the SNR was fixed at 5 dB in the single interferer scenario.

As expected, the bilateral benefit was smaller when multiple noise sources were present (see Figure 3.8). There is ample evidence in the binaural hearing literature suggesting that when both devices are available, CI listeners can benefit from the so-called head-shadow effect which occurs mainly when speech and noise are spatially separated (e.g., see Shaw, 1974; Hawley et al., 2004; Litovsky et al., 2009). For three out of five bilateral subjects tested, the scores indicated that the signal from the noise source coming from the right side was severely attenuated prior to reaching the left ear. In the case where there was only one interferer (steady-state noise) on the right (+90°), a better SNR was obtained at the left ear compared to the right. The individual listener was able to selectively focus on the left ear placed contralateral to the competing noise source to improve speech recognition.

By comparing the scores obtained with the 2M2-BEAM processing strategy against the baseline bilateral unprocessed condition, we observe that there is a marginal improvement in speech intelligibility. Still, the results indicate that the subjects tested were not able to significantly benefit from the use of the binaural beamformer over their daily strategy. We attribute that to the presence of reverberation. As depicted in Figure 3.9, this effect was more prominent in the multiple interferer scenario. This relatively small improvement noted can be in part attributed to the fact that the performance of beamforming-based processing strategies has been known to degrade sharply in reverberant conditions (e.g., see Greenberg and Zurek, 1992; Hamacher *et al.*, 1997; Kompis and Dillier, 1994; van Hoesel and Clark, 1995; Ricketts and Henry, 2002; Spriet *et al.*, 2007).

In the study by van Hoesel and Clark (1995), beamforming attenuated spatially separated noise by 20 dB in anechoic settings but only by 3 dB in highly reverberant settings. More recently, another study by Lockwood *et al.* (2004) focusing on measuring the effects of reverberation demonstrated that the performance of beamforming algorithms is decreased by substantial amounts as the reverberation time of the room increases. Ricketts and Henry (2002) measured speech intelligibility in hearing aid users under moderate (0.3 sec) and severe levels of reverberation (0.9 sec) and showed that the benefit due to beamforming decreases, and in fact, it almost entirely disappears near reflective surfaces or highly reverberant environments. Spriet *et al.* (2007) reached a similar outcome and concluded that reverberation had a detrimental effect on the performance of the two-microphone BEAM strategy. In accordance with the aforementioned research studies, the data from this study suggest that the binaural beamformer (2M2-BEAM) strategy cannot enhance the speech recognition ability of CI users considerably under the reverberant listening conditions tested.

In contrast, the 4M-SESS strategy yielded a considerable benefit in all conditions tested and for all five subjects. As evidenced from the scores plotted in Figure 3.8 and Figure 3.9, the observed benefit over the subjects' daily strategy ranged from 30 percentage points when multiple interferers were present to around 50 percentage points for the case where only a single interferer emanated from a single location in space. This improvement in performance with the proposed 4M-SESS strategy was maintained even in the challenging condition where three noise sources were present and were asymmetrically located across the two hemifields. In addition, the overall benefit after processing with the 4M-SESS processing strategy was significantly higher than the benefit received when processing with the binaural 2M2-BEAM noise reduction strategy. In both the single and multi-noise source scenarios, the 4M-SESS strategy employing two contralateral microphone signals led to a substantially increased performance, especially when the speech and the noise source were spatially separated. The observed improvement in performance with the 4M-SESS strategy can be attributed to having a more reliable reference signal, even in situations where the noise sources are asymmetrically located across the two hemifields.

In essence, combining spatial information from all four microphones forms a better representation of the reference signal, leading to a better target segregation than that made available with only a monaural input or two binaural inputs. Having four microphones working simultaneously enables us to form a better estimate of the speech component and the generated noise reference.

Subsequently, speech intelligibility performance is considerably better. One may interpret this as introducing or enhancing the better-ear advantage into the noise reduction processing strategy. The implementation of the 4M-SESS processing strategy requires access to a single processor driving both cochlear implants. Signals arriving from the left and right sides need to be captured synchronously in order to be further processed together. Co-ordinated binaural stimulation through the 4M-SESS strategy by utilizing input signals from both the left and right CI devices offers a much greater potential for noise reduction than using signals from just a single CI or two independent CI devices. As wireless communication of audio signals is quickly becoming possible between the two sides, it may be advantageous to have the two devices in a bilateral fitting communicate with each other (e.g., see Schum, 2008). In the future, a more intelligent signal transmission scheme that would enable a signal exchange between the two cochlear implant processors could also be developed and tested further.

CHAPTER 4

Conclusions and future challenges

This booklet investigated one of the most commercially attractive applications of BSS, which is the simultaneous recovery of signals inside a reverberant (naturally echoing) environment, using two (or more) microphones. The goal of this booklet in the lecture series was to provide insight on recent advances in algorithms, which are ideally suited for blind signal separation of convolutive mixtures. In the convolutive mixing scenario paradigm, each microphone captures not only the direct contributions from each original sound source, but also several reflected copies of the original signals at different propagation delays.

We started by presenting the necessary background material regarding the blind signal separation framework (Chapter 1). We then provided a detailed mathematical description of the BSS problem and presented a thorough review of the generative models for both instantaneous and convolutive mixtures. Next, we discussed the most important theoretical aspects that underpin the development of convolutive blind signal separation strategies, and we also reviewed some of the most promising BSS algorithms for speech processing applications (Chapter 2). We then focused on the application of these signal processing strategies to noise reduction for hearing-impaired listeners (Chapter 3).

Such algorithms can boost speech intelligibility in noisy conditions and further enhance the ability of hearing-impaired individuals to communicate better in noise. We extensively described noise reduction processing strategies based on BSS, which can exploit an array of two or more microphones (Chapter 3). Speech intelligibility studies with hearing-impaired listeners, who were aided by BSS-enabled prosthetic devices (cochlear implants), were conducted in non-optimal listening scenarios. Experimental results revealed a considerable benefit in all conditions tested and for all subjects and therefore clearly demonstrated the potential of incorporating BSS strategies in future auditory prosthetic and hearing aid devices.

The concept is indeed novel, as well as highly practical. Blind signal separation is a well studied research topic, however, numerous issues remain largely unresolved. Some of the most important open issues and challenges are listed below:

Real-time operation — All algorithms discussed here were implemented in their off-line forms. The ability of the algorithms to perform in real-time is most appropriate for their application to auditory prostheses and hearing aids, as discussed in Chapter 3. For example, a practically useful BSS strategy should perform reasonably well even inside a moderately-to-severely reverberant environment yielding at least 5 dB to almost 10 dB improvement after processing. In addition, this level of performance needs to be reached within 1-2 seconds for the strategy to be deemed as usable and practical. More research and further testing needs to be carried out in order to determine how much delay can incur between the incoming sound and the processed sound reaching the listener's ears. Also, a more in depth study focusing on the trade-off between separation performance and computational complexity is necessary.

Accommodating more signals than sensors — In general, at least the same number of sources as the number of mixtures is required. Hence, in scenarios where the number of sources is larger than the number of sensors, it is almost impossible to design a linear separation system that extracts all source signals. Therefore, it remains an open problem how to extract useful source estimates in such situations. Such problems have significant practical interest, as it is often the case that mixtures of many source signals are measured by only a few sensors in a microphone array.

Speed of convergence and tracking behavior — The majority of adaptive BSS algorithms, which employ blind criteria for training purposes suffer from slow convergence properties. Such slow convergence properties impose fundamental limitations on performance and therefore prohibit the use of these algorithms in practical applications. In addition, slow convergence may prevent adequate tracking of the unknown system conditions, especially in situations where the parameters of the unknown mixing system are changing with respect to time due to source (speaker) movements.

We conclude this booklet by suggesting a number of potential avenues for further research on the subject of BSS:

Hybrid signal separation — Blind signal separation and adaptive beamforming have a similar goal. Both techniques attempt to enhance signals of interest by optimizing a filter structure, which can reduce interference by essentially forming a null directivity pattern towards the undesired sound source. Therefore, the respective advantages of these two techniques can be combined, while their weaknesses can be compensated with each other. A hybrid strategy is expected to greatly enhance computational efficiency and overall performance.

Adaptive filter order — Normally, most of the reverberant energy in the acoustic impulse response is concentrated in the first 10–20 msecs. Exploiting this fact, the overall efficiency of convolutive BSS algorithms could be improved further, by essentially implementing an adaptive filter order building up scheme, which would initially use a short unmixing FIR filter and then would switch to a longer one, as the algorithm approaches convergence. The adoption of such scheme could speed up algorithm convergence considerably.

Selective-tap filtering — Incorporate a partial-updating mechanism in the error gradient of the update equation. In the adopted CBSS strategies we can, in principle, update only a selected portion of the adaptive filters. The algorithms will then steer all the computational resources to the filter taps having the largest magnitude gradient components on the error surface. In this case, the algorithms will theoretically require only a small number of updates at each iteration and therefore will operate with a substantially reduced computational complexity.

Non-parametric source modeling — In the source modeling aspect of the BSS problem, rather than using a parametric nonlinear score function to approximate the source distributions of the data, an interesting topic of further research would be to allow the statistics of the (recovered) data themselves to fully determine the nonlinear transform that best describes them. In that respect, a promising way forward would be to resort to a non-parametric PDF estimator, such as the kernel or Parzen density estimation technique.

Bibliography

Aiazzi B., Alparone L. and Baronti S. (1999), "Estimation based on entropy matching for generalized Gaussian PDF modeling," *IEEE Signal Processing Letters*, vol. 6, no. 6, pp. 138–140. DOI: 10.1109/97.763145 30

Alain C., Reinke K., He Y., Wang C. and Lobaugh N. (2005), "Hearing two things at once: neurophysiological indices of speech segregation and identification," *Journal of Cognitive Neuroscience*, vol. 17, no. 5, pp. 811–818. DOI: 10.1162/0898929053747621 1

Amari S.-I., Cichocki A. and Yang H. H. (1996), "A new learning algorithm for blind signal separation," *Advances in Neural Information Processing Systems*, MA: MIT Press, vol. 8, pp. 757–763. 28

Amari S.-I., (1999), "Natural gradient learning for over- and under-complete bases in ICA," *Neural Computation*, vol. 11, no. 8, pp. 1875–1883. DOI: 10.1162/089976699300015990 5

Araki S., Sawada H. and Makino S. (2007), "K-means based underdetermined blind speech separation," in *Blind Speech Separation* (Eds. Makino S., Lee T.-W. and Sawada H.), pp. 243–270. DOI: 10.1007/978-1-4020-6479-1_9 5, 42

Arbogast T. L., Mason C. R. and Kidd G. (2002), "The effect of spatial separation on informational and energetic masking of speech," *Journal of the Acoustical Society of America*, vol. 112, no. 5, pp. 2086–2098. DOI: 10.1121/1.1510141 62

Back A. D. and Weigend A. S. (1997), "A first application of independent component analysis to extracting structure from stock returns," *International Journal of Neural Systems*, vol. 8, no. 4, pp. 473–484. DOI: 10.1142/S0129065797000458 3

Bell A. J. and Sejnowski T. J. (1995), "An information maximization approach to blind separation and blind deconvolution," *Neural Computation*, vol. 7, no. 6, pp. 1129–1159. DOI: 10.1162/neco.1995.7.6.1129 24, 25, 27, 28, 30, 38

Berkhout A. J., De Vries D. and Boone M. M. (1980), "A new method to acquire impulse responses in concert hall," *Journal of the Acoustical Society of America*, vol. 68, no. 1, pp. 179–183. DOI: 10.1121/1.384618 11

Blauert J., Brueggen M., Bronkhorst A. W., Drullman R., Reynaud G., Pellieux L., Krebber W. and Sottek R. (1998), "The AUDIS catalog of human HRTFs," *Journal of the Acoustical Society of America*, vol. 103, pp. 3082. DOI: 10.1121/1.422910 58

76   BIBLIOGRAPHY

Bronkhorst A. W. (2000), "The cocktail party phenomenon: A review of research on speech intelligibility in multiple-talker conditions *Acustica*, vol. 86, pp. 117–128. 1

Brown G. J and Cooke M. (1994), "Computational auditory scene snalysis," *Computer Speech and Language*, vol. 8, no. 4, pp. 297–336. DOI: 10.1006/csla.1994.1016 1

Cardoso J.-F. (1989), "Source separation using higher-order moments", in *Proc. IEEE International Conference on Acoustics, Speech and Signal Processing*, Glasgow, Scotland, May 23–26, pp. 2109–2112. 7

Cardoso J.-F. (1997), "Infomax and maximum likelihood for source separation," *IEEE Signal Processing Letters*, vol. 4, no. 4, pp. 112–114. DOI: 10.1109/97.566704 26

Cardoso J.-F. (1998), "Blind signal separation: Statistical principles," *Proceedings of the IEEE*, vol. 86, no. 10, pp. 2009–2025. DOI: 10.1109/5.720250 6

Cardoso J.-F. and Laheld B. H. (1996), "Equivariant adaptive source separation," *IEEE Transactions on Signal Processing*, vol. 44, no. 12, pp. 3017–3030. DOI: 10.1109/78.553476 7, 29

Chaumette E., Comon P. and Muller D. (1993), "ICA-Based techniques for radiating sources estimation: Application to airport surveillance," *IEE Proceedings*, vol. 140, no. 6, pp. 395–401. 2

Cherry E. C. (1953), "Some experiments on the recognition of speech with one and with two ears," *Journal of the Acoustical Society of America*, vol. 25, no. 5, pp. 975–979. DOI: 10.1121/1.1907229 1

Cherry E. C. and Taylor W. K. (1954), "Some further experiments upon the recognition of speech with one and with two ears," *Journal of the Acoustical Society of America*, vol. 26, no. 4, pp. 554–559. DOI: 10.1121/1.1907373 1

Cheung Y.-M. and Xu L. (2001), "Independent component ordering in ICA time series analysis," *Neurocomputing*, vol. 41, no. 1–4, pp. 145–152. DOI: 10.1016/S0925-2312(00)00358-1 3

Choi S., Cichocki A. and Amari S.-I. (2000), "Flexible independent component analysis," *Journal of VLSI Signal Processing*, vol. 26, no. 1, pp. 25–38. DOI: 10.1023/A:1008135131269 32

Chung K. (2004), "Challenges and recent developments in hearing aids. Part I. Speech understanding in noise, microphone technologies and noise reduction algorithms," *Trends In Amplification*, vol. 8, no. 3, pp. 8–124. DOI: 10.1177/108471380400800302 51

Chung K., Zeng F.-G. and Acker K. N. (2006), "Effects of directional microphone and adaptive multichannel noise reduction algorithm on cochlear implant performance," *Journal of the Acoustical Society of America*, vol. 120, no. 4, pp. 2216–2227. DOI: 10.1121/1.2258500 51

Cichocki A. and Amari S.-I. (2002), *Adaptive Blind Signal and Image Processing: Learning Algorithms and Applications*. New York: John Wiley & Sons, Inc. DOI: 10.1002/0470845899 5, 17

Clark G. (2003), *Cochlear Implants: Fundamentals and Applications.* New York: Springer-Verlag. 49

Comon P. (1994), "Independent component analysis: A new concept?" *Signal Processing*, vol. 36, no. 3, pp. 287–314. DOI: 10.1016/0165-1684(94)90029-9 2, 4, 5, 7, 24

Davenport W. B. (1952), "An experimental study of speech-wave probability distributions," *Journal of the Acoustical Society of America*, vol. 24, no. 4, pp. 390–399. DOI: 10.1121/1.1906909 30

Deco G. and Obradovic D. (1996), *An Information Theoretic Approach to Neural Computing.* New York: Springer-Berlin. 24

Do M. N. and Vetterli M. (2002), "Wavelet-based texture retrieval using generalized Gaussian density and Kullback-Leibler distance," *IEEE Transactions on Image Processing*, vol. 11, no. 2, pp. 146–158. DOI: 10.1109/83.982822 30

Douglas S. C. and Haykin S. (2000), "Relationships between blind deconvolution and blind source separation," in *Unsupervised Adaptive Filtering, Volume II: Blind Deconvolution* (Ed. Haykin S.), pp. 113–145. 38, 40

Douglas S. C., Sawada H. and Makino S. (2005), "Natural gradient multichannel blind deconvolution and speech separation using causal FIR filters," *IEEE Transactions on Speech and Audio Processing*, vol. 13, no.1, pp. 92–104. DOI: 10.1109/TSA.2004.838538 38, 40

Freyman R. L., Balakrishnan U. and Helfer K. S. (2001), "Spatial release from informational masking in speech recognition," *Journal of the Acoustical Society of America*, vol. 109, no. 5, pp. 2112–2122. DOI: 10.1121/1.1354984 62

Gates G. A. and Miyamoto R. T. (2003), "Cochlear implants," *New England Journal of Medicine*, vol. 349, pp. 421–423. DOI: 10.1056/NEJMp038107 49

Godard D. N. (1980), "Self-recovering equalization and carrier tracking in two-dimensional data communication systems," *IEEE Transactions on Communications*, vol. 28, no. 11, pp. 1867–1875. DOI: 10.1109/TCOM.1980.1094608 3

Greenberg Z. E. and Zurek P. M. (1992), "Evaluation of an adaptive beamforming method for hearing aids," *Journal of the Acoustical Society of America*, vol. 91, no. 3, pp. 1662–1676. DOI: 10.1121/1.402446 52, 54, 68

Grbić N., Tao X.-J., Nordholm S. E.and Claesson I. (2001), "Blind signal separation using over-complete subband representation," *IEEE Transaction on Speech and Audio Processing*, vol. 9, no. 5, pp. 524–533. DOI: 10.1109/89.928917 40

Griffiths L. J. and Jim C. W. (1982), "An alternative approach to linearly constrained adaptive beamforming," *IEEE Transactions on Antennas and Propagation*, vol. AP-30, pp. 27–34. DOI: 10.1109/TAP.1982.1142739 51

Hamacher V., Doering W., Mauer G., Fleischmann H. and Hennecke J. (1997), "Evaluation of noise reduction systems for cochlear implant users in different acoustic environments," *American Journal of Otology*, vol. 18, no. 6, pp. 46–49. 51, 52, 68

Hawkins D. B. and Yacullo W. S. (1984), "Signal-to-noise ratio advantage of binaural hearing aids and directional microphones under different levels of reverberation," *Journal of Speech and Hearing Disorders*, vol. 49, pp. 278–286. 67

Hawley M. L., Litovsky R. Y. and Culling J. F. (2004), "The benefit of binaural hearing in a cocktail party: Effect of location and type of interferer," *Journal of the Acoustical Society of America*, vol. 115, no. 2, pp. 833–843. DOI: 10.1121/1.1639908 61, 67

Haykin S. (1996), *Adaptive Filter Theory* (New Jersey: Prentice Hall). 54

Hochberg I., Boothroyd A., Weiss M. and Hellman S. (1992), "Effects of noise and noise suppression on speech perception for cochlear implant users,," *Ear and Hearing*, vol. 13, no.4, pp. 263–271. 50

House W. F. and Berliner K. I. (1991), "Cochlear implants: From idea to clinical practice," in *Cochlear Implants: A Practical Guide* (Ed. Cooper H.), pp. 9–33. 49

Hu Y., Loizou P. C. and Kasturi, K. (2007), "Use of a sigmoidal-shaped function for noise attenuation in cochlear implants," *Journal of the Acoustical Society of America*, vol. 122, no. 4, pp. 128–134. DOI: 10.1121/1.2772401 50

Hyvärinen A. (1999), "Gaussian moments for noisy independent component analysis," *IEEE Signal Processing Letters*, vol. 6, no. 6, pp. 145–147. DOI: 10.1109/97.763148 14

Hyvärinen A., Karhunen J. and Oja E. (2001), *Independent Component Analysis*. New York: John Wiley & Sons, Inc. DOI: 10.1002/0471221317 6, 8

Ikeda S., Murata N. and Ziehe A. (2001), "An approach to blind source separation based on temporal structure of speech signals," *Neurocomputing*, vol. 41, no. 1–4, pp. 1–24. DOI: 10.1016/S0925-2312(00)00345-3 41

Ikram M. Z. and Morgan D. R. (2005), "Permutation inconsistency in blind speech separation: Investigation and solutions," *IEEE Transactions on Speech and Audio Processing*, vol. 13, no. 1, pp. 1–13. DOI: 10.1109/TSA.2004.834441 36, 37, 38, 42

IEEE (1969), "IEEE recommended practice speech quality measurements," *IEEE Transactions on Audio and Electroacoustics*, vol. 17, no. 3, pp. 225–246. 58

ITU-T (2001), "Perceptual evaluation of speech quality (PESQ), an objective method for end-to-end speech quality assessment of narrow-band telephone networks and speech coders," ITU-T Recommendation P. 862, 2001. 45

Jackson C. W. and Turnbull, A. (2004), "Impact of deafness on family life: A review of the literature,"*Topics in Early Childhood Special Education*, vol. 24, no.1, pp. 15–29. DOI: 10.1177/02711214040240010201 48

Joho M. (2000), *A Systematic Approach to Adaptive Algorithms for Multichannel System Identification, Inverse Modeling and Blind Identification*. Ph.D. Thesis, ETH Zurich. 38

Joshi R. L. and Fisher T. R. (1995), "Comparison of generalized Gaussian and Laplacian modeling in DCT image coding," *IEEE Signal Processing Letters*, vol. 2, no. 5, pp. 81–82. DOI: 10.1109/97.386283 30

Jutten C. and Hérault J. (1991), "Blind separation of sources, Part I: An adaptive algorithm based on neuromimetic architecture," *Signal Processing*, vol. 24, no. 1, pp. 1–10. DOI: 10.1016/0165-1684(91)90079-X 1, 7

Kandel E. R., Schwartz J. H and Jessel T. M. (1991), *The Principles of Neural Science*. New York: Prentice Hall. 24

Karhunen J., Hyvärinen A., Vigário R., Hurri J. and Oja E. (1997), "Applications of neural blind separation to signal and image processing," in *Proc. IEEE International Conference on Acoustics, Speech and Signal Processing*, Munich, Germany, April 21–24, pp. 131–134. DOI: 10.1109/ICASSP.1997.599569 3

Kates J. M. (2008), *Digital Hearing Aids*. Boston: Plural Publishing, Inc. 48

Kawamoto M., Matsuoka K. and Ohnishi N. (1998), "A method of blind separation for convolved non-stationary signals," *Neurocomputing*, vol. 22, no. 1–3, pp. 157–171. DOI: 10.1016/S0925-2312(98)00055-1 36

Kinsler L. E., Frey A. R., Coppens A. B. and Sanders J. V. (2000), *Fundamentals of Acoustics*. United Kingdom: John Wiley & Sons, Inc. 11

Kokkinakis K. and Nandi A. K. (2005), "Exponent parameter estimation for generalized Gaussian probability density functions with application to speech modeling," *Signal Processing*, vol. 85, no. 9, pp. 1852–1858. DOI: 10.1016/j.sigpro.2005.02.017 30, 32

Kokkinakis K. and Nandi A. K. (2006), "Multichannel blind deconvolution for source separation in convolutive mixtures of speech," *IEEE Transactions on Speech and Audio Processing*, vol. 14, no. 1, pp. 200–213. DOI: 10.1109/TSA.2005.854109 38, 39, 40

Kokkinakis K. and Loizou P. C. (2007), "Subband-based blind signal processing for source separation in convolutive mixtures of speech," in *Proc. IEEE International Conference on Acoustics, Speech and Signal Processing*, Honolulu, Hawaii, April 15–20, pp. 917–920. DOI: 10.1109/ICASSP.2007.367220 40

Kokkinakis K. and Loizou P. C. (2008), "Using blind source separation techniques to improve speech recognition in bilateral cochlear implant patients," *Journal of the Acoustical Society of America*, vol. 123, no. 4, pp. 2379–2390. DOI: 10.1121/1.2839887 2, 50, 53, 56

Kokkinakis K. and Loizou P. C. (2009), "Selective-tap blind dereverberation for two-microphone enhancement of reverberant speech," *IEEE Signal Processing Letters*, vol. 16, no. 11, pp. 961–964. DOI: 10.1109/LSP.2009.2027658 21

Kokkinakis K. and Loizou P. C. (2010), "Multi-microphone adaptive noise reduction strategies for co-ordinated stimulation in bilateral cochlear implant devices," *Journal of the Acoustical Society of America*, in press. 50, 53

Kompis, M., and Dillier, N. (1994), "Noise reduction for hearing aids: Combining directional microphones with an adaptive beamformer," *Journal of the Acoustical Society of America*, vol. 96, no. 3, pp. 1134–1143. DOI: 10.1121/1.410204 52, 68

Kullback S. (1959), *Information Theory and Statistics*. United Kingdom: John Wiley & Sons, Inc. 24

Kuttruf H. (1991), *Room Acoustics*. New York: Elsevier Science Publishers Ltd. 11, 12

Lambert R. H. (1996), *Multichannel Blind Deconvolution: FIR Matrix Algebra and Separation of Multipath Mixtures*. Ph.D. Thesis, University of Southern California, Los Angeles. 15, 39

Lambert R. H. and Bell A. J. (1997), "Blind separation of multiple speakers in a multipath environment," *Proc. IEEE International Conference on Acoustics, Speech and Signal Processing*, Munich, Germany, April 21–24, pp. 423–426. DOI: 10.1109/ICASSP.1997.599665 39

Lee T.-W., Bell A. J. and Lambert R. H. (1997), "Blind separation of delayed and convolved Sources," *Advances in Neural Information Processing Systems*, MA: MIT Press, vol. 9, pp. 758–764. 39

Lee T.-W., Lewicki M. S., Girolami M. and Sejnowski T. J. (1999), "Blind source separation of more sources than mixtures using overcomplete representations," *IEEE Signal Processing Letters*, vol. 6, no. 4, pp. 87–90. DOI: 10.1109/97.752062 5

Lee J.-H., Oh S.-H and Lee S.-Y. (2008), "Binaural semi-blind dereverberation of noisy convoluted speech signals," *Neurocomputing*, vol. 72, no. 1–3, pp. 636–642. DOI: 10.1016/j.neucom.2008.07.005 40

Lewicki M. S. (1994), "Bayesian modeling and classification of neural signals," *Neural Computation*, vol. 6, no. 5, pp. 1005–1030. DOI: 10.1162/neco.1994.6.5.1005 3

Lipkin M. and Williams M. E. (1986), "Presbycusis and communication," *Journal Journal of General Internal Medicine*, vol. 1, no. 6 , pp. 399–401. DOI: 10.1007/BF02596426 48

Litovsky R., Parkinson A., Arcaroli J., Peters R., Lake J., Johnstone P. and Yu G. (2004), "Bilateral cochlear implants in adults and children," *Archives of Otolaryngology Head and Neck Surgery*, vol. 130, no. 5, pp. 648–655. DOI: 10.1001/archotol.130.5.648 49

Litovsky R., Parkinson A. and Arcaroli J. (2009), " Spatial hearing and speech intelligibility in bilateral cochlear implant users," *Ear and Hearing*, vol. 30, no. 4, pp. 419–431. DOI: 10.1097/AUD.0b013e3181a165be 49, 67

Lockwood M. E., Jones D. L., Bilger R. C., Lansing C. R., O'Brien W. D., Wheeler B. C. and Feng A. S. (2004), "Performance of time- and frequency-domain binaural beamformers based on recorded signals from real rooms," *Journal of the Acoustical Society of America*, vol. 115, no. 1, pp. 379–391. DOI: 10.1121/1.1624064 68

Loeb G. E. (1990), "Cochlear prosthetics," *Annual Review of Neuroscience*, vol. 13, pp. 357–371. DOI: 10.1146/annurev.ne.13.030190.002041 49

Loizou P. C. (1998), "Mimicking the human ear," *IEEE Signal Processing Magazine*, vol. 15, pp. 101–130. DOI: 10.1109/79.708543 49

Loizou P. C., Lobo A. and Hu Y. (2005), "Subspace algorithms for noise reduction in cochlear implants," *Journal of the Acoustical Society of America*, vol. 118, no. 5, pp. 2791–2793. DOI: 10.1121/1.2065847 50

Loizou P. C. (2006), "Speech processing in vocoder-centric cochlear implants," in *Cochlear and Brainstem Implants* (Ed. Moller A.), pp. 109–143. 50

Loizou P. C., Hu Y., Litovsky R., Yu G., Peters R. Lake, J., and Roland P. (2009), "Speech recognition by bilateral cochlear implant users in a cocktail party setting," *Journal of the Acoustical Society of America*, vol. 125, no., pp. 372–383. DOI: 10.1121/1.3036175 50

Makeig S., Jung T.-P., Bell A. J. and Sejnowski T. J. (1996), "Independent component analysis in elecroengephalographic data," *Advances in Neural Information Processing Systems*, MA: MIT Press, vol. 8, pp. 145–151. 3

Manmontri U. and Naylor P. A. (2008), "A class of Frobenius norm-based algorithms using penalty term and natural gradient for blind signal separation," *IEEE Transactions on Speech and Audio Processing*, vol. 16, no. 6, pp. 1181–1193. DOI: 10.1109/TASL.2008.2001388 36, 37, 38

Matsuoka K., Ohya M. and Kawamoto M. (1995), "A neural net for blind separation of non-stationary signals," *Neural Networks*, vol. 8, no. 3, pp. 411–419. DOI: 10.1016/0893-6080(94)00083-X 36

Matsuoka K. and Nakashima S. (2001), "Minimal distortion principle for blind source separation," In *Proc. International Symposium on Independent Component Analysis and Blind Signal Separation*, San Diego, California, USA, December 9–13, pp. 722–727. DOI: 10.1109/SICE.2002.1195729 41

Mitianoudis N. and Davies M. E. (2003), "Audio source separation of convolutive mixtures," *IEEE Transactions on Speech and Audio Processing*, vol. 11, no. 5, pp. 489–497. DOI: 10.1109/TSA.2003.815820 42

Miyoshi M. and Kaneda Y. (1988), "Inverse filtering of room acoustics," *IEEE Transactions on Acoustics, Speech and Signal Processing*, vol. 36, no. 2, pp. 145–152. DOI: 10.1109/29.1509 13

Moore B. C. J. (2003), "Speech processing for the hearing-impaired: successes, and implications for speech mechanisms," *Speech Communication*, vol. 41, no. 1, pp. 81–91. DOI: 10.1016/S0167-6393(02)00095-X 48

Moore B. C. J. (2007), *Cochlear Hearing Loss: Physiological, Psychological and Technical Issues.* Chichester: John Wiley & Sons, Inc. DOI: 10.1002/9780470987889 47, 50

Moreau E. and Pesquet J.-C. (1997), "Generalized contrasts for multichannel blind deconvolution of linear systems," *IEEE Signal Processing Letters*, vol. 4, no. 6, pp. 182–183. DOI: 10.1109/97.586043 14

Nadal J.-P. and Parga N. (1994), "Non-linear neurons in the low noise limit: A factorial code maximizes information transfer," *Network: Computation in Neural Systems*, vol. 5, no. 4, pp. 565–581. DOI: 10.1088/0954-898X/5/4/008 24

Nandi A. K. (1999), *Blind Estimation Using Higher-Order Statistics.* Boston: Kluwer Academic Publishers. 7

Neely S. T. and Allen J. B. (1979), "Invertibility of a room impulse response," *Journal of the Acoustical Society of America*, vol. 66, no. 1, pp. 165–169. DOI: 10.1121/1.383069 13

Neubauer R. (2001), "Estimation of reverberation time in rectangular rooms with nonuniformly distributed absorption using a modified Fitzroy equation," *Building Acoustics*, vol. 8, pp. 115–137. DOI: 10.1260/1351010011501786 13

Omologo M., Svaizer P. and Matassoni M. (1998), "Environmental conditions and acoustic transduction in hands-free speech recognition," *Speech Communication*, vol. 25, no. 9, pp. 75–95. DOI: 10.1016/S0167-6393(98)00030-2 2

Oppenheim A. V and Schafer R. W (1989), *Disrcete-Time Signal Processing.* New Jersey: Prentice Hall. 16

Palmer C. V. and Ortmann A. (2005), "Hearing loss and hearing aids," *Neurologic Clinics*, vol. 23, no. 3, pp. 901–918. DOI: 10.1016/j.ncl.2005.03.002 48

Papoulis A. (1991), *Probability, Random Variables, and Stochastic Processes*. New York: McGraw-Hill. 25, 27

Parra L. and Spence C. (2000), "Convolutive blind separation of non-stationary sources," *IEEE Transactions on Speech and Audio Processing*, vol. 8, no. 3, pp. 320–327. DOI: 10.1109/89.841214 15, 36, 37, 38, 41, 42

Parra L. and Alvino C. (2002), "Geometric source separation: Merging convolutive source separation with geometric beamforming," *IEEE Transactions on Speech and Audio Processing*, vol. 10, no. 6, pp. 352–362. DOI: 10.1109/TSA.2002.803443 38, 42

Patrick J. F., Busby P. A. and Gibson P. J. (2006), "The development of the Nucleus Freedom cochlear implant system," *Trends in Amplification*, vol. 10, no. 4, pp. 175–200. DOI: 10.1177/1084713806296386 54

Pearlmutter B. and Parra L. (1996), "Maximum likelihood blind source separation: A context sensitive generalization of ICA," *Advances in Neural Information Processing Systems*, MA: MIT Press, vol. 8, pp. 613–619. 26

Pedersen M. S., Larsen J., Kjems U., Parra L. (2007), "A survey of convolutive blind source separation methods", in *Springer Handbook of Speech Processing* (Eds. Benesty J., Sondhi M. M. and Huang Y.), pp. 1065–1094. 36, 41, 42

Peters R. W., Moore B. C. J. and Baer T. (1998), "Speech reception thresholds in noise with and without spectral and temporal dips for hearing-impaired and normally hearing people" *Journal of the Acoustical Society of America*, vol. 103, no. 1, pp. 577–587. DOI: 10.1121/1.421128 50

Plomp R., (1994), "Noise, amplification, and compression: Considerations of three main issues in hearing aid design," *Ear and Hearing*, vol. 15, no. 1, pp. 2–12. 50

Pope K. J. and Bogner R. E. (1996a), "Blind signal separation I: Linear instantaneous combinations," *Digital Signal Processing*, vol. 6, no. 2, pp. 5–16. DOI: 10.1006/dspr.1996.0002 3

Pope K. J. and Bogner R. E. (1996b), "Blind signal separation II: Linear convolutive combinations," *Digital Signal Processing*, vol. 6, no. 3, pp. 17–28. DOI: 10.1006/dspr.1996.0003 11

Richards D. L. (1964), "Statistical properties of speech signals," *Proceedings of the IEE*, vol. 111, no. 5, pp. 941–949. 30

Ricketts, T. A., and Henry, P. (2002), "Evaluation of an adaptive directional-microphone hearing aid," *International Journal of Audiology*, vol. 41, no. 2 , pp. 100–112. DOI: 10.3109/14992020209090400 68

Saberi K., Dostal L., Sadralodabai T., Bull V. and Perrot D. R. (1991), "Free-field release from masking," *Journal of the Acoustical Society of America*, vol. 90, no. 3, pp. 1355–1370. DOI: 10.1121/1.401927 61

Schobben D. W. E, Torkkola K. and Smaragdis P. (1999), "Evaluation of blind signal separation methods," In *Proc. First International Workshop on Independent Component Analysis and Blind Signal Separation*, Aussois, France, January 11-15, pp. 261–266. 13

Schobben D. W. E. and Sommen P. C. W. (2002), "A frequency domain blind signal separation method based on decorrelation," *IEEE Transactions on Signal Processing*, vol. 50, no. 8, pp. 1855–1865. DOI: 10.1109/TSP.2002.800417 36, 41, 42

Schroeder M. R. (1965), "New method for measuring the reverberation time," *Journal of the Acoustical Society of America*, vol. 37, pp. 409–412. DOI: 10.1121/1.1909343

Schum D. J (2008), "Communication between hearing aids," *Advance for Audiologists*, vol. 10, pp. 44–49. 69

Sharifi K. and Leon-Garcia A. (1995), "Estimation of shape parameter for generalized Gaussian distributions in subband decompositions of video," *IEEE Transactions on Circuits and Systems for Video Technology*, vol. 5, no. 1, pp. 52–56. DOI: 10.1109/76.350779 30

Shaw E. A. (1974), "Transformation of sound pressure level from free field to the eardrum in the horizontal plane," *Journal of the Acoustical Society of America*, vol. 56, pp. 1848–1861. DOI: 10.1121/1.1903522 67

Smaragdis P. (1998), "Blind separation of convolved mixtures in the frequency domain," *Neurocomputing*, vol. 22, no. 1–3, pp. 21–34. DOI: 10.1016/S0925-2312(98)00047-2 15, 38, 41, 42

Smits C., Kramer, S. E. and Houtgast T.(2006), "Speech reception thresholds in noise and self-reported hearing disability in a general adult population," *Ear and Hearing*, vol. 27, no. 5, pp. 538–549. DOI: 10.1097/01.aud.0000233917.72551.cf 50

Spahr A. J. and Dorman M. F. (2004), "Performance of subjects fit with the Advanced Bionics CII and Nucleus 3G cochlear implant devices," *Archives of Otolaryngology Head and Neck Surgery*, vol. 130, no. 5, pp. 624–628. DOI: 10.1001/archotol.130.5.624 50

Spriet A., Van Deun L., Eftaxiadis K., Laneau J., Moonen M., Van Dijk B., Van Wieringen A. and Wouters J. (2007), "Speech understanding in background noise with the two-microphone adaptive beamformer BEAM in the Nucleus Freedom cochlear implant system," *Ear and Hearing*, vol. 28, no. 1, pp. 62–72. 51, 52, 54, 68

Stickney G. S., Zeng F.-G., Litovsky R. and Assmann P. F. (2004), "Cochlear implant speech recognition with speech maskers," *Journal of the Acoustical Society of America*, vol. 116, no. 2, pp. 1081–1091. DOI: 10.1121/1.1772399 50, 61

Talwar S., Viberg M. and Paulraj A. (1996), "Blind separation of synchronous co-channel digital signals using an antenna array. Part I: Algorithms," *IEEE Transactions on Signal Processing*, vol. 44, no. 5, pp. 1184–1197. DOI: 10.1109/78.502331 3

Thirion N., Mars J. and Boelle J.-L. (1996), "Separation of seismic signals: A new concept based on a blind algorithm," in *Proc. Eighth European Signal Processing Conference*, Trieste, Italy, September 10–13, pp. 85–88. 3

Tonazzini A., Bedini L. and Salerno E. (2004), "Independent component analysis for document restoration," *International Journal on Document Analysis and Recognition*, vol. 7, no. 1, pp. 17–27. DOI: 10.1007/s10032-004-0121-8 3

Tong L., Inouye Y. and Liu R. (1993), "Waveform-preserving blind estimation of multiple independent sources," *IEEE Transactions on Signal Processing,* vol. 41, no. 7, pp. 2461–2470. DOI: 10.1109/78.224254 7

Torkkola K. (1999), "Blind Separation for Audio Signals–Are We There Yet?" In *Proc. First International Workshop on Independent Component Analysis and Blind Signal Separation*, Aussois, France, January 11–15, pp. 239–244. 25

Tugnait J. K. (1994), "Blind estimation of digital communication channel impulse response," *IEEE Transactions on Communications*, vol. 43, no. 2, pp. 1606–1616. DOI: 10.1109/TCOMM.1994.582855 3

Tyler R. S., Gantz B. J., Rubinstein J. T., Wilson B. S., Parkinson A. J., Wolaver A., Preece J. P., Witt S. and Lowder M. W. (2002), "Three-month results with bilateral cochlear implants," *Ear and Hearing*, vol. 23, no. 1, pp. 80–89. 61

van den Bogaert T., Doclo S., Wouters J. and Moonen M. (2009), "Speech enhancement with multichannel Wiener filter techniques in multimicrophone binaural hearing aids," *Journal of the Acoustical Society of America*, vol. 125, no. 1, pp. 360–-371. DOI: 10.1121/1.3023069 52, 60

van Hoesel R. J. M. and Clark G. M. (1995), "Evaluation of a portable two-microphone adaptive beamforming speech processor with cochlear implant patients" *Journal of the Acoustical Society of America*, vol. 97, no. 4, pp. 2498–2503. DOI: 10.1121/1.411970 51, 68

van Hoesel R. J. M. and Tyler R. S. (2003), "Speech perception, localization, and lateralization with binaural cochlear implants," *Journal of the Acoustical Society of America*, vol. 113, no. 3, pp. 1617–1630. 49, 57, 61

Vandali A. E., Whitford L. A., Plant K. L. and Clark G. M. (2000), "Speech perception as a function of electrical stimulation rate: Using the Nucleus 24 Cochlear implant system," *Ear and Hearing*, vol. 21, no. 6, pp. 608–624. DOI: 10.1097/00003446-200012000-00008 57

Visser E., Otsuka M. and Lee T.-W. (2003), "Spatio-temporal speech enhancement scheme for robust speech recognition in noisy environments," *Speech Communication*, vol. 41, no. 2–3, pp. 393–407. DOI: 10.1016/S0167-6393(03)00010-4 2

Walden A. T. (1985), "Non-Gaussian feflectivity, entropy and deconvolution," *Geophysics*, vol. 12, pp. 2862–2888. DOI: 10.1190/1.1441905 3

Wang D. L. and Brown G. J. (2006), *Computational Auditory Scene Analysis: Principles, Algorithms and Applications*.   New Jersey: Wiley, Hoboken. 1

Wang W., Sanei S. and Chambers J. A. (2005), Penalty function-based joint diagonalization approach for convolutive blind separation of nonstationary sources," *IEEE Transactions on Signal Processing*, vol. 53, no. 5, pp. 1654–1669. 37, 38

Weiss M. (1993), "Effects of noise and noise reduction processing on the operation of the Nucleus 22 cochlear implant processor," *Journal of Rehabilitation Research and Development*, vol. 30, no. 1, pp. 117–128. 50

Weintraub M. (1985), *A Theory and Computational Model of Auditoty Monoraul Sound Separation*. Ph.D. Thesis, Stanford University, San Francisco. 1

Wouters J. and van den Berghe J. (2001), "Speech recognition in noise for cochlear implantees with a two microphone monaural adaptive noise reduction system," *Ear and Hearing*, vol. 22, no. 5, pp. 420–430. DOI: 10.1097/00003446-200110000-00006 51, 54

Yang, L.-P., and Fu, Q.-J. (2005), "Spectral subtraction-based speech enhancement for cochlear implant patients in background noise," *Journal of the Acoustical Society of America*, vol. 117, no. 3, pp. 1001–1004. DOI: 10.1121/1.1852873 50

Zarzoso V. and Nandi A. K. (2001), "Non-invasive fetal electrocardiogram extraction: Blind separation versus adaptive noise cancellation," *IEEE Transactions on Biomedical Engineering*, vol. 48, no. 1, pp. 12–18. DOI: 10.1109/10.900244 3

Authors' Biographies

KOSTAS KOKKINAKIS

Kostas Kokkinakis graduated from the University of Sheffield, United Kingdom, with the B.S. degree in Electronics, Control and Systems Engineering in 2000. He then received the M.S. degree in Microelectronics and Signal Processing from the University of London and the Ph.D. degree in Electronics from the University of Liverpool, in 2001 and 2005, respectively. Dr. Kokkinakis is currently a Research Assistant Professor in the Erik Jonsson School of Engineering and Computer Science at the University of Texas at Dallas, Richardson, TX, working on multi-microphone signal processing strategies for speech enhancement. The focus of his research work is on statistical signal processing, psychoacoustics, speech modeling and noise reduction. His research interests lie mainly in the development of blind signal separation strategies and their application to cochlear implant devices. His research is supported by funding from the National Institute on Deafness and other Communication Disorders of the National Institutes of Health.

PHILIPOS C. LOIZOU

Philipos C. Loizou received the B.S., M.S., and Ph.D. degrees, all in Electrical Engineering, from Arizona State University (ASU), Tempe, AZ, in 1989, 1991, and 1995, respectively. From 1995 to 1996, he was a Postdoctoral Fellow in the Department of Speech and Hearing Science at ASU, working on research related to cochlear implants. He was an Assistant Professor at the University of Arkansas at Little Rock from 1996 to 1999. He is now a Professor and holder of the Cecil and Ida Green Chair in the Department of Electrical Engineering, University of Texas at Dallas, Richardson, TX. His research interests are in the areas of signal processing, speech processing and cochlear implants. Dr. Loizou is currently working on the development of novel speech processing algorithms that will aid people with hearing impairment, and in particularly, people wearing cochlear implants. He is author of the book *Speech Enhancement: Theory and Practice* (CRC, 2007) and co-author of the textbook *An Interactive Approach to Signals and Systems Laboratory* (National Instruments, 2008). Dr. Loizou is a Fellow of the Acoustical Society of America and a member of the Speech Technical Committee of the IEEE Signal Processing Society. His research is supported by funding from the National Institute on Deafness and other Communication Disorders of the National Institutes of Health.

Printed in the United States
by Baker & Taylor Publisher Services